A
FIELD GUIDE
TO LIES

A
FIELD GUIDE
TO LIES

Critical Thinking
in the Information Age

Daniel J. Levitin

ALLEN
LANE

ALLEN LANE

an imprint of Penguin Canada, a division of Penguin Random House Canada Limited

Canada • USA • UK • Ireland • Australia • New Zealand • India • South Africa • China

Published in Allen Lane hardcover by Penguin Canada, 2016
Simultaneously published in the United States by Dutton, an imprint of Penguin Random House LLC

All art courtesy of the author unless otherwise noted.
Images on pp. 7, 80, 174, 185, 240 © 2016 by Dan Piraro, used by permission.
Image on p. 25 © 2016 by Alex Tabarrok, used by permission.
Image on p. 27 was drawn by the author, based on a figure under Creative Commons license
appearing on www.betterposters.blogspot.com.
Image on p. 49 © 2016 by Tyler Vigen, used by permission.
Image on p. 207 was redrawn by the author with permission, based on a figure found at
AutismSpeaks.org.
Image on p. 247 is public domain and provided courtesy of Harrison Prosper.

www.penguinrandomhouse.ca

LIBRARY AND ARCHIVES CANADA CATALOGUING IN PUBLICATION

Levithin, Daniel J., author
A field guide to lies / Daniel J. Levithin.

ISBN 978-0-670-06994-1 (hardcover)
ISBN 978-0-14-319628-0 (electronic)

1. Statistics--Evaluation. 2. Science news--Evaluation.
3. Electronic information resources--Evaluation. 4. Information
literacy. 5. Critical thinking. 6. Reasoning. 7. Trust.
I. Title.

HA29.L49 2016   001.4'22   C2015-908758-9

Cover design by Evan Gaffney
Cover image by Alamy Stock Photo

Printed and bound in the United States of America

10 9 8 7 6 5 4 3 2 1

Penguin
Random
House
ALLEN
LANE

To Shari,
whose inquisitive mind made me a better thinker

CONTENTS

Introduction: Thinking, Critically   ix

PART ONE: EVALUATING NUMBERS

Plausibility   3

Fun with Averages   11

Axis Shenanigans   26

Hijinks with How Numbers Are Reported   43

How Numbers Are Collected   75

Probabilities   97

PART TWO: EVALUATING WORDS

How Do We Know?   123

Identifying Expertise   129

Overlooked, Undervalued Alternative Explanations   152

Counterknowledge   168

CONTENTS

PART THREE: EVALUATING THE WORLD

How Science Works   181

Logical Fallacies   198

Knowing What You Don't Know   211

Bayesian Thinking in Science and in Court   216

Four Case Studies   222

Conclusion: Discovering Your Own   251

Appendix: Application of Bayes's Rule   255

Glossary   257

Notes   263

Acknowledgments   283

Index   285

INTRODUCTION

Thinking, Critically

This is a book about how to spot problems with the facts you encounter, problems that may lead you to draw the wrong conclusions. Sometimes the people giving you the facts are hoping you'll draw the wrong conclusion; sometimes they don't know the difference themselves. Today, information is available nearly instantaneously, but it is becoming increasingly hard to tell what's true and what's not, to sift through the various claims we hear and to recognize when they contain misinformation, pseudo-facts, distortions, and outright lies.

There are many ways that we can be led astray by fast-talking, loose-writing purveyors of information. Here, I've grouped them into two categories, and they make the first two parts of this book: numerical and verbal. The first includes mishandled statistics and graphs; the second includes faulty arguments. In both parts, I include the steps we can take to better evaluate news, statements, and reports. The last part of the book addresses what underlies our ability to determine if something is true or false: the scientific method. It grapples with the limits of what we can and cannot know, including what we know right now and don't know just yet, and includes some applications of logical thinking.

It is easy to lie with statistics and graphs because few people take the time to look under the hood and see how they work. I aim to fix that. Recognizing faulty arguments can help you to evaluate whether a chain of reasoning leads to a valid conclusion or not. Related to this is infoliteracy—recognizing that there are hierarchies in source quality, that pseudo-facts can easily masquerade as facts, and biases can distort the information we are being asked to consider, leading us to faulty conclusions.

You might object and say, "But it's not my job to evaluate statistics critically. Newspapers, bloggers, the government, Wikipedia, etc., should be doing that for us." Yes, they should, but they don't always. We—each of us—need to think critically and carefully about the numbers and words we encounter if we want to be successful at work, at play, and in making the most of our lives. This means checking the numbers, the reasoning, and the sources for plausibility and rigor. It means examining them as best as we can before we repeat them or use them to form an opinion. We want to avoid the extremes of gullibly accepting every claim we encounter or cynically rejecting every one. Critical thinking doesn't mean we disparage everything, it means that we try to distinguish between claims with evidence and those without.

Sometimes the evidence consists of numbers and we have to ask, "Where did those numbers come from? How were they collected?" Sometimes the numbers are ridiculous, but it takes some reflection to see it. Sometimes claims seem reasonable, but come from a source that lacks credibility, like a person who reports having witnessed a crime but wasn't actually there. This book can help you to avoid learning a whole lot of things that aren't so. And catch some lying weasels in their tracks.

We've created more human-made information in the last five

years than in all of human history before them. Unfortunately, found alongside things that are true is an enormous number of things that are not, in websites, videos, books, and on social media. This is not just a new problem. Misinformation has been a fixture of human life for thousands of years, and was documented in biblical times and classical Greece. The unique problem we face today is that misinformation has proliferated; it is devilishly entwined on the Internet with real information, making the two difficult to separate. And misinformation is promiscuous—it consorts with people of all social and educational classes, and turns up in places you don't expect it to. It propagates as one person passes it on to another and another, as Twitter, Facebook, Snapchat, and other social media grab hold of it and spread it around the world; the misinformation can take hold and become well known, and suddenly a whole lot of people are believing things that aren't so.

PART ONE

EVALUATING NUMBERS

It ain't what you don't know that gets you into trouble.
It's what you know for sure that just ain't so.

—MARK TWAIN

Plausibility

Statistics, because they are numbers, appear to us to be cold, hard facts. It seems that they represent facts given to us by nature and it's just a matter of finding them. But it's important to remember that *people* gather statistics. People choose what to count, how to go about counting, which of the resulting numbers they will share with us, and which words they will use to describe and interpret those numbers. Statistics are not facts. They are interpretations. And your interpretation may be just as good as, or better than, that of the person reporting them to you.

Sometimes, the numbers are simply wrong, and it's often easiest to start out by conducting some quick plausibility checks. After that, even if the numbers pass plausibility, three kinds of errors can lead you to believe things that aren't so: how the numbers were collected, how they were interpreted, and how they were presented graphically.

In your head or on the back of an envelope you can quickly determine whether a claim is plausible (most of the time). Don't just accept a claim at face value; work through it a bit.

When conducting plausibility checks, we don't care about the exact numbers. That might seem counterintuitive, but precision isn't important here. We can use common sense to reckon a lot of these: If Bert tells you that a crystal wineglass fell off a table and hit a thick carpet without breaking, that seems plausible. If Ernie says it fell off the top of a forty-story building and hit the pavement without breaking, that's not plausible. Your real-world knowledge, observations acquired over a lifetime, tells you so. Similarly, if someone says they are two hundred years old, or that they can consistently beat the roulette wheel in Vegas, or that they can run forty miles an hour, these are not plausible claims.

What would you do with this claim?

> In the thirty-five years since marijuana laws stopped being enforced in California, the number of marijuana smokers has doubled every year.

Plausible? Where do we start? Let's assume there was only one marijuana smoker in California thirty-five years ago, a very conservative estimate (there were half a million marijuana arrests nationwide in 1982). Doubling that number every year for thirty-five years would yield more than 17 billion—larger than the population of the entire world. (Try it yourself and you'll see that doubling every year for twenty-one years gets you to over a million: 1; 2; 4; 8; 16; 32; 64; 128; 256; 512; 1024; 2048; 4096; 8192; 16,384; 32,768; 65,536; 131,072; 262,144; 524,288; 1,048,576.) This claim isn't just implausible, then, it's impossible. Unfortunately, many people have trouble thinking clearly about numbers because they're intimidated by them. But as you see, nothing here requires more than elementary school arithmetic and some reasonable assumptions.

Here's another. You've just taken on a position as a telemarketer, where agents telephone unsuspecting (and no doubt irritated) prospects. Your boss, trying to motivate you, claims:

> Our best salesperson made 1,000 sales a day.

Is this plausible? Try dialing a phone number yourself—the fastest you can probably do it is five seconds. Allow another five seconds for the phone to ring. Now let's assume that every call ends in a sale—clearly this isn't realistic, but let's give every advantage to this claim to see if it works out. Figure a minimum of ten seconds to make a pitch and have it accepted, then forty seconds to get the buyer's credit card number and address. That's one call per minute (5 + 5 + 10 + 40 = 60 seconds), or 60 sales in an hour, or 480 sales in a very hectic eight-hour workday with no breaks. The 1,000 just isn't plausible, allowing even the most optimistic estimates.

Some claims are more difficult to evaluate. Here's a headline from *Time* magazine in 2013:

> More people have cell phones than toilets.

What to do with this? We can consider the number of people in the developing world who lack plumbing and the observation that many people in prosperous countries have more than one cell phone. The claim seems *plausible*—that doesn't mean we should accept it, just that we can't reject it out of hand as being ridiculous; we'll have to use other techniques to evaluate the claim, but it passes the plausibility test.

Sometimes you can't easily evaluate a claim without doing a bit of research on your own. Yes, newspapers and websites really ought to be doing this for you, but they don't always, and that's how runaway statistics take hold. A widely reported statistic some years ago was this:

> In the U.S., 150,000 girls and young women die of anorexia each year.

Okay—let's check its plausibility. We have to do some digging. According to the U.S. Centers for Disease Control, the annual number of deaths *from all causes* for girls and women between the ages of fifteen and twenty-four is about 8,500. Add in women from twenty-five to forty-four and you still only get 55,000. The anorexia deaths in one year cannot be three times the number of *all* deaths.

In an article in *Science*, Louis Pollack and Hans Weiss reported that since the formation of the Communication Satellite Corp.,

> The cost of a telephone call has decreased by 12,000 percent.

If a cost decreases by 100 percent, it drops to zero (no matter what the initial cost was). If a cost decreases by 200 percent, someone is paying *you* the same amount you used to pay *them* for you to take the product. A decrease of 100 percent is very rare; one of 12,000 percent seems wildly unlikely. An article in the peer-reviewed *Journal of Management Development* claimed a 200 percent reduction in customer complaints following a new customer care strategy.

Author Dan Keppel even titled his book *Get What You Pay For: Save 200% on Stocks, Mutual Funds, Every Financial Need.* He has an MBA. He should know better.

Of course, you have to apply percentages to the same baseline in order for them to be equivalent. A 50 percent reduction in salary cannot be restored by increasing your new, lower salary by 50 percent, because the baselines have shifted. If you were getting $1,000/ week and took a 50 percent reduction in pay, to $500, a 50 percent increase in that pay only brings you to $750.

Percentages seem so simple and incorruptible, but they are often confusing. If interest rates rise from 3 percent to 4 percent, that is an increase of 1 percentage point, or 33 percent (because the 1 percent rise is taken against the baseline of 3, so 1/3 = .33). If interest rates fall from 4 percent to 3 percent, that is a decrease of 1 percentage point, but not a decrease of 33 percent—it's a decrease of 25 percent (because the 1 percentage point drop is now taken against the baseline of 4). Researchers and journalists are not always scrupulous about making this distinction between percentage point and percentages clear, but you should be.

The *New York Times* reported on the closing of a Connecticut textile mill and its move to Virginia due to high employment costs. The *Times* reported that employment costs, "wages, worker's compensation and unemployment insurance—are 20 times higher in Connecticut than in Virginia." Is this plausible? If it were true, you'd think that there would be a mass migration of companies out of Connecticut and into Virginia—not just this one mill—and that you would have heard of it by now. In fact, this was not true and the *Times* had to issue a correction. How did this happen? The reporter simply misread a company report. One cost, unemployment insurance, was in fact twenty times higher in Connecticut than in Virginia, but when factored in with other costs, total employment costs were really only 1.3 times higher in Connecticut, not 20 times higher. The reporter did not have training in business administration and we shouldn't expect her to. To catch these kinds of errors requires taking a step back and thinking for ourselves—which anyone can do (and she and her editors should have done).

New Jersey adopted legislation that denied additional benefits to mothers who have children while already on welfare. Some legislators believed that women were having babies in New Jersey simply to increase the amount of their monthly welfare checks. Within two months, legislators were declaring the "family cap" law a great success because births had already fallen by 16 percent. According to the *New York Times:*

> After only two months, the state released numbers suggesting that births to welfare mothers had already fallen by 16

percent, and officials began congratulating themselves on their overnight success.

Note that they're not counting pregnancies, but births. What's wrong here? Because it takes nine months for a pregnancy to come to term, any effect in the first two months cannot be attributed to the law itself but is probably due to normal fluctuations in the birth rate (birth rates are known to be seasonal).

Even so, there were other problems with this report that can't be caught with plausibility checks:

> . . . over time, that 16 percent drop dwindled to about 10 percent as the state belatedly became aware of births that had not been reported earlier. It appeared that many mothers saw no reason to report the new births since their welfare benefits were not being increased.

This is an example of a problem in the way statistics were collected—we're not actually surveying all the people that we think we are. Some errors in reasoning are sometimes harder to see coming than others, but we get better with practice. To start, let's look at a basic, often misused tool.

The pie chart is an easy way to visualize percentages—how the different parts of a whole are allocated. You might want to know what percentage of a school district's budget is spent on things like salaries, instructional materials, and maintenance. Or you might want to know what percentage of the money spent on instructional materials goes toward math, science, language arts, athletics, music, and so on. The cardinal rule of a pie chart is that the percentages have to add up to 100. Think about an actual pie—if there are nine people who each want an equal-sized piece, you can't cut it into eight. After you've reached the end of the pie, that's all there is. Still, this didn't stop Fox News from publishing this pie chart:

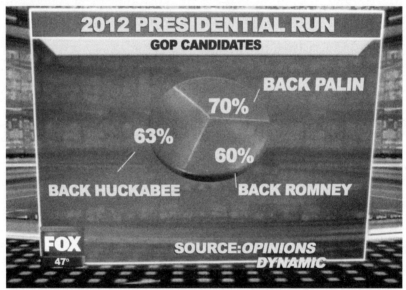

First rule of pie charts: The percentages have to add up to 100. (*Fox News, 2010*)

You can imagine how something like this could happen. Voters are given the option to report that they support more than one candidate. But then, the results shouldn't be presented as a pie chart.

Fun with Averages

An average can be a helpful summary statistic, even easier to digest than a pie chart, allowing us to characterize a very large amount of information with a single number. We might want to know the average wealth of the people in a room to know whether our fundraisers or sales managers will benefit from meeting with them. Or we might want to know the average price of gas to estimate how much it will cost to drive from Vancouver to Banff. But averages can be deceptively complex.

There are three ways of calculating an average, and they often yield different numbers, so people with statistical acumen usually avoid the word *average* in favor of the more precise terms *mean, median,* and *mode.* We don't say "mean average" or "median average" or simply just "average"—we say *mean, median,* or *mode.* In some cases, these will be identical, but in many they are not. If you see the word *average* all by itself, it's usually indicating the mean, but you can't be certain.

The mean is the most commonly used of the three and is calculated by adding up all the observations or reports you have and dividing by the number of observations or reports. For example, the average wealth of the people in a room is simply the total wealth divided by the number of people. If the room has ten people whose net worth is $100,000 each, the room has a total net worth of $1 million, and you can figure the mean without having to pull out a calculator: It is $100,000. If a different room has ten people whose net worth varies from $50,000 to $150,000 each, but totals $1 million, the mean is still $100,000 (because we simply take the total $1 million and divide by the ten people, regardless of what any individual makes).

The median is the middle number in a set of numbers (statisticians call this set a "distribution"): Half the observations are above it and half are below. Remember, the point of an average is to be able to represent a whole lot of data with a single number. The median does a better job of this when some of your observations are very, very different from the majority of them, what statisticians call *outliers*.

If we visit a room with nine people, suppose eight of them have a net worth of near $100,000 and one person is on the verge of bankruptcy with a net worth of negative $500,000, owing to his debts. Here's the makeup of the room:

Person 1: −$500,000
Person 2: $96,000
Person 3: $97,000
Person 4: $99,000
Person 5: $100,000
Person 6: $101,000
Person 7: $101,000
Person 8: $101,000
Person 9: $104,000

Now we take the sum and obtain a total of $299,000. Divide by the total number of observations, nine, and the mean is $33,222 per person. But the mean doesn't seem to do a very good job of characterizing the room. It suggests that your fund-raiser might not want to visit these people, when it's really only one odd person, one outlier, bringing down the average. This is the problem with the mean: It is sensitive to outliers.

The median here would be $100,000: Four people make less than that amount, and four people make more. The mode is $101,000, the number that appears more often than the others. Both the median and the mode are more helpful in this particular example.

There are many ways that averages can be used to manipulate what you want others to see in your data.

Let's suppose that you and two friends founded a small start-up company with five employees. It's the end of the year and you want to report your finances to your employees, so that they can feel good about all the long hours and cold pizzas they've eaten, and so that you can attract investors. Let's say that four employees— programmers—each earned $70,000 per year, and one employee—a receptionist/office manager—earned $50,000 per year. That's an average (mean) employee salary of $66,000 per year (4 × $70,000) + (1 × $50,000), divided by 5. You and your two friends each took home $100,000 per year in salary. Your payroll costs were therefore (4 × $70,000) + (1 × $50,000) + (3 × $100,000) = $630,000. Now, let's say your company brought in $210,000 in profits and you divided it equally among you and your co-founders as bonuses, giving you $100,000 + $70,000 each. How are you going to report this?

You could say:

Average salary of employees: $66,000
Average salary + profits of owners: $170,000

This is true but probably doesn't look good to anyone except you and your mom. If your employees get wind of this, they may feel undercompensated. Potential investors may feel that the founders are overcompensated. So instead, you could report this:

> Average salary of employees: $66,000
> Average salary of owners: $100,000
> Profits: $210,000

That looks better to potential investors. And you can just leave out the fact that you divided the profits among the owners, and leave out that last line—that part about the profits—when reporting things to your employees. The four programmers are each going to think they're very highly valued, because they're making more than the average. Your poor receptionist won't be so happy, but she no doubt knew already that the programmers make more than she does.

Now suppose you are feeling overworked and want to persuade your two partners, who don't know much about critical thinking, that you need to hire more employees. You could do what many companies do, and report the "profits per employee" by dividing the $210,000 profit among the five employees:

> Average salary of employees: $66,000
> Average salary of owners: $100,000
> Annual profits per employee: $42,000

Now you can claim that 64 percent of the salaries you pay to employees (42,000/66,000) comes back to you in profits, meaning you end up only having to pay 36 percent of their salaries after all those profits roll in. Of course, there is nothing in these figures to suggest that adding an employee will increase the profits—your profits may not be at all a function of how many employees there are—but for someone who is not thinking critically, this sounds like a compelling reason to hire more employees.

Finally, what if you want to claim that you are an unusually just and fair employer and that the difference between what you take in profits and what your employees earn is actually quite reasonable? Take the $210,000 in profits and distribute $150,000 of it as salary bonuses to you and your partners, saving the other $60,000 to report as "profits." This time, compute the average salary but include you and your partners in it with the salary bonuses.

> Average salary: $97,500
> Average profit of owners: $20,000

Now for some real fun:

> Total salary costs plus bonuses: $840,000
> Salaries: $780,000
> Profits: $60,000

That looks quite reasonable now, doesn't it? Of the $840,000 available for salaries and profits, only $60,000 or 7 percent went into owners' profits. Your employees will think you above reproach—who would begrudge a company owner from taking 7 percent? And it's actually not even that high—the 7 percent is divided among the three company owners to 2.3 percent each. Hardly worth complaining about!

You can do even better than this. Suppose in your first year of operation, you had only part-time employees, earning $40,000 per year. By year two, you had only full-time employees, earning the $66,000 mentioned above. You can honestly claim that average employee earnings went up 65 percent. What a great employer you are! But here you are glossing over the fact that you are comparing part-time with full-time. You would not be the first: U.S. Steel did it back in the 1940s.

In criminal trials, the way the information is presented—the framing—profoundly affects jurors' conclusions about guilt. Although they are mathematically equivalent, testifying that "the probability the suspect would match the blood drops if he were not their source is only 0.1 percent" (one in a thousand) turns out to be far more persuasive than saying "one in a thousand people in Houston would also match the blood drops."

Averages are often used to express outcomes, such as "one in X marriages ends in divorce." But that doesn't mean that statistic will apply on your street, in your bridge club, or to anyone you know. It might or might not—it's a nationwide average, and there might be certain *vulnerability factors* that help to predict who will and who will not divorce.

Similarly, you may read that one out of every five children born is Chinese. You note that the Swedish family down the street already has four children and the mother is expecting another child. This does not mean she's about to give birth to a Chinese baby—the one out of five children is on average, across all births in the world, not the births restricted to a particular house or particular neighborhood or even particular country.

Be careful of averages and how they're applied. One way that they can fool you is if the average combines samples from disparate populations. This can lead to absurd observations such as:

On average, humans have one testicle.

This example illustrates the difference between mean, median, and mode. Because there are slightly more women than men in the world, the median and mode are both zero, while the mean is close to one (perhaps 0.98 or so).

Also be careful to remember that the average doesn't tell you anything about the range. The average annual temperature in Death Valley, California, is a comfortable 77 degrees F (25 degrees C). But the range can kill you, with temperatures ranging from 15 degrees to 134 degrees on record.

Or . . . I could tell you that the *average* wealth of a hundred people in a room is a whopping $350 million. You might think this is the place to unleash a hundred of your best salespeople. But the room could have Mark Zuckerberg (net worth $35 billion) and ninety-nine people who are indigent. The average can smear across differences that are important.

Another thing to watch out for in averages is the *bimodal distribution*. Remember, the *mode* is the value that occurs most often. In many biological, physical, and social datasets, the distribution has two or more peaks—that is, two or more values that appear more than the others.

Bimodal Distribution

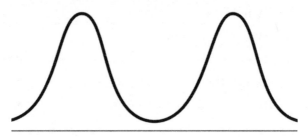

For example, a graph like this might show the amount of money spent on lunches in a week (x-axis) and how many people spent that amount (y-axis). Imagine that you've got two different groups of people in your survey, children (left hump—they're buying school lunches) and business executives (right hump—they're going to fancy restaurants). The mean and median here could be a number somewhere right between the two, and would not tell us very much about what's really going on—in fact, the mean and median in many cases are amounts that nobody spends. A graph like this is often a clue that there is heterogeneity in your sample, or that you are comparing apples and oranges. Better here is to report that it's a bimodal distribution and report the two modes. Better yet, subdivide the group into two groups and provide statistics for each.

But be careful drawing conclusions about individuals and groups based on averages. The pitfalls here are so common that they have names: the ecological fallacy and the exception fallacy. The ecological fallacy occurs when we make inferences about an individual based on aggregate data (such as a group mean), and the exception fallacy occurs when we make inferences about a group based on knowledge of a few exceptional individuals.

For example, imagine two small towns, each with only one hundred people. Town A has ninety-nine people earning $80,000 a year, and one super-wealthy person who struck oil on her property, earning $5,000,000 a year. Town B has fifty people earning $100,000 a year and fifty people earning $140,000. The mean income of Town A is $129,200 and the mean income of Town B is $120,000. Although Town A has a higher mean income, in ninety-nine out of one hundred cases, any individual you select randomly from Town B will have a higher income than an individual selected randomly from Town A. The ecological fallacy is thinking that if you select someone at random from the group with the higher mean, that individual is likely to have a higher income. The neat thing is, in the examples above, that it's not just the *mean* that is higher in Town B but also the *median* and the *mode*. (It doesn't always work out that way.)

As another example, it has been suggested that wealthy individuals are more likely to vote Republican, but evidence shows that the wealthier states tend to vote Democratic. The wealth of those wealthier states may be skewed by a small percentage of super-wealthy individuals. During the 2004 U.S. presidential election, the Republican candidate, George W. Bush, won the fifteen poorest states, and the Democratic candidate, John Kerry, won nine of the eleven wealthiest states. However, 62 percent of those with annual incomes over $200,000 voted for Bush, whereas only 36 percent of voters with annual incomes of $15,000 or less voted for Bush.

As an example of the exception fallacy, you may have read that Volvos are among the most reliable automobiles and so you decide to buy one. On your way to the dealership, you pass a Volvo mechanic and find a parking lot full of Volvos in need of repair. If you change your mind about buying a Volvo based on seeing this, you're using a relatively small number of exceptional cases to form an inference about the entire group. No one was claiming that Volvos never need repair, only that they're less likely to in the aggregate. (Hence the ubiquitous cautionary note in advertising that "individual performance may vary.") Note also that you're being unduly influenced by this in another way: The one place that Volvos needing repair will be is at a Volvo mechanic. Your "base rate" has shifted, and you cannot consider this a random sample.

Now that you're an expert on averages, you shouldn't fall for the famous misunderstanding that people tended not to live as long a hundred years ago as they do today. You've probably read that life expectancy has steadily increased in modern times. For those born in 1850, the average life expectancy for males and females was thirty-eight and forty years respectively, and for those born in 1990 it is seventy-two and seventy-nine. There's a tendency to think, then, that in the 1800s there just weren't that many fifty- and sixty-year-olds walking around because people didn't live that long. But in fact, people did live that long—it's just that infant and childhood mortality was so high that it skewed the average. If you could make it past twenty, you could live a long life back then. Indeed, in 1850 a fifty-year-old white female could expect to live to be 73.5, and a sixty-year-old could expect to live to be seventy-seven. Life expectancy has certainly increased for fifty- and sixty-year-olds today, by about ten years compared to 1850, largely due to better health care.

But as with the examples above of a room full of people with wildly different incomes, the changing averages for life expectancy at birth over the last 175 years reflect significant differences in the two samples: There were many more infant deaths back then pulling down the average.

Here is a brain-twister: The average child usually doesn't come from the average family. Why? Because of shifting baselines. (I'm using "average" in this discussion instead of "mean" out of respect for a wonderful paper on this topic by James Jenkins and Terrell Tuten, who used it in their title.)

Now, suppose you read that the average number of children per family in a suburban community is three. You might conclude then that the average child must have two siblings. But this would be wrong. This same logical problem applies if we ask whether the average college student attends the average-sized college, if the average employee earns the average salary, or if the average tree comes from the average forest. What?

All these cases involve a shift of the baseline, or sample group we're studying. When we calculate the average number of children per family, we're sampling families. A very large family and a small family each count as one family, of course. When we calculate the average (mean) number of siblings, we're sampling children. Each child in the large family gets counted once, so that the number of siblings each of them has weighs heavily on the average for sibling number. In other words, a family with ten children counts only one time in the average *family* statistic, but counts ten times in the average *number of siblings* statistic.

Suppose in one neighborhood of this hypothetical community

there are thirty families. Four families have no children, six families have one child, nine families have two children, and eleven families have six children. The average number of children per family is three, because ninety (the total number of children) gets divided by thirty (the total number of families).

But let's look at the average number of siblings. The mistake people make is thinking that if the average family has three children, then each child must have two siblings on average. But in the one-child families, each of the six children has zero siblings. In the two-child families, each of the eighteen children has one sibling. In the six-child families each of the sixty-six children has five siblings. Among the 90 children, there are 348 siblings. So although the average *child* comes from a family with three children, there are 348 siblings divided among 90 children, or an average of nearly four siblings per child.

	Families	# Children/ Family	Total # Children	Siblings
	4	0	0	0
	6	1	6	0
	9	2	18	18
	11	6	66	330
Totals	30		90	348

Average children per family: 3.0
Average siblings per child: 3.9

FUN WITH AVERAGES

4 Families with 0 children

6 Families with 1 child — 6 children with 0 siblings

9 Families with 2 children — 18 children with 1 sibling

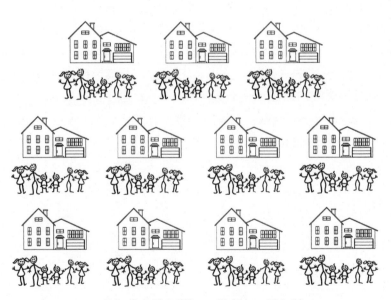

11 Families with 6 children — 66 children with 5 siblings

Consider now college size. There are many very large colleges in the United States (such as Ohio State and Arizona State) with student enrollment of more than 50,000. There are also many small colleges, with student enrollment under 3,000 (such as Kenyon College and Williams College). If we count up *schools*, we might find that the average-sized college has 10,000 students. But if we count up students, we'll find that the average student goes to a college with greater than 30,000 students. This is because, when counting students, we'll get many more data points from the large schools. Similarly, the average person doesn't live in the average city, and the average golfer doesn't shoot the average round (the total strokes over eighteen holes).

These examples involve a shift of baseline, or denominator. Consider another involving the kind of skewed distribution we looked at earlier with child mortality: The average investor does not earn the average return. In one study, the average return on a $100 investment held for thirty years was $760, or 7 percent per year. But 9 percent of the investors lost money, and a whopping 69 percent failed to reach the average return. This is because the average was skewed by a few people who made much greater than the average— in the figure below, the *mean* is pulled to the right by those lucky investors who made a fortune.

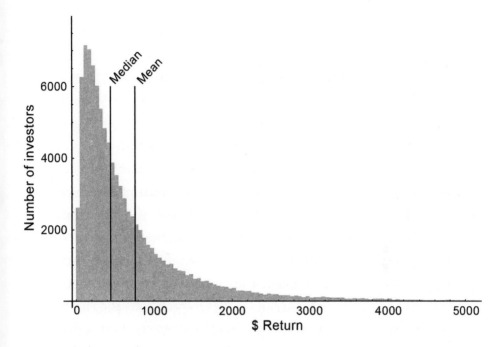

Payoff outcomes for return on a $100 investment over thirty years. Note that most people make less than the mean return, and a lucky few make more than five times the mean return.

Axis Shenanigans

The human brain did not evolve to process large amounts of numerical data presented as text; instead, our eyes look for patterns in data that are visually displayed. The most accurate but least interpretable form of data presentation is to make a table, showing every single value. But it is difficult or impossible for most people to detect patterns and trends in such data, and so we rely on graphs and charts. Graphs come in two broad types: Either they represent every data point visually (as in a scatter plot) or they implement a form of data reduction in which we summarize the data, looking, for example, only at means or medians.

There are many ways that graphs can be used to manipulate, distort, and misrepresent data. The careful consumer of information will avoid being drawn in by them.

Unlabeled Axes

The most fundamental way to lie with a statistical graph is to not label the axes. If your axes aren't labeled, you can draw or plot anything you want! Here is an example from a poster presented at a conference by a student researcher, which looked like this (I've redrawn it here):

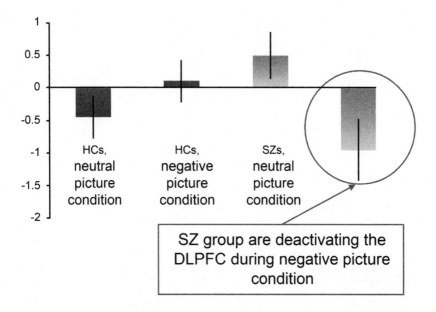

SZ group are deactivating the DLPFC during negative picture condition

What does all that mean? From the text on the poster itself (though not on this graph), we know that the researchers are studying brain activations in patients with schizophrenia (SZ). What are HCs? We aren't told, but from the context—they're being compared with SZ—we might assume that it means "healthy controls." Now, there do appear to be differences between the HCs and the SZs, but, hmmm . . . the y-axis has numbers, but . . . the units could be anything! What are we looking at? Scores on a test, levels of brain activations, number of brain regions activated? Number of Jell-O brand pudding cups they've eaten, or number of Johnny Depp movies they've seen in the last six weeks? (To be fair, the researchers subsequently published their findings in a peer-reviewed journal, and corrected this error after a website pointed out the oversight.)

In the next example, gross sales of a publishing company are plotted, excluding data from Kickstarter campaigns.

Gross Sales Excluding Kickstarter

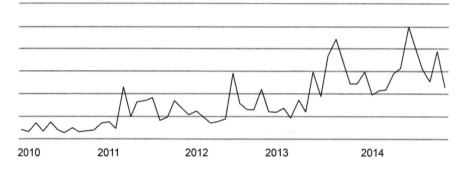

As in the previous example, but this time with the x-axis, we have numbers but we're not told what they are. In this case, it's probably self-evident: We assume that the 2010, 2011, etc., refer to calendar or fiscal years of operation, and the fact that the lines are jagged between the years suggests that the data are being tracked monthly (but without proper labeling we can only assume). The y-axis is completely missing, so we don't know what is being measured (is it units sold or dollars?), and we don't know what each horizontal line represents. The graph could be depicting an increase of sales from 50 cents a year to $5 a year, or from 50 million to 500 million units. Not to worry—a helpful narrative accompanied this graph: "It's been another great year." I guess we'll have to take their word for it.

Truncated Vertical Axis

A well-designed graph clearly shows you the relevant end points of a continuum. This is especially important if you're documenting some actual or projected change in a quantity, and you want your readers to draw the right conclusions. If you're representing crime rate, deaths, births, income, or any quantity that could take on a value of zero, then zero should be the minimum point on your graph. But if your aim is to create panic or outrage, start your y-axis somewhere near the lowest value you're plotting—this will emphasize the difference you're trying to highlight, because the eye is drawn to the size of the difference as shown on the graph, and the actual size of the difference is obscured.

In 2012, Fox News broadcast the following graph to show what would happen if the Bush tax cuts were allowed to expire:

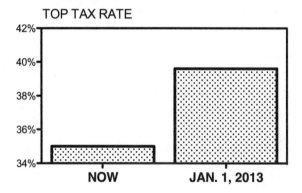

IF BUSH TAX CUTS EXPIRE

TOP TAX RATE

The graph gives the visual impression that taxes would increase by a large amount: The right-hand bar is six times the height of the left-hand bar. Who wants their taxes to go up by a factor of six? Viewers who are number-phobic, or in a hurry, may not take the time to examine the axis to see that the actual difference is between a tax rate of 35 percent and one of 39.6 percent. That is, if the cuts expire, taxes will only increase 13 percent, not the 600 percent that is pictured (the 4.6 percentage point increase is 13 percent of 35 percent).

If the y-axis started at zero, the 13 percent would be apparent visually:

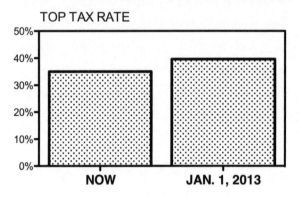

Discontinuity in Vertical or Horizontal Axis

Imagine a city where crime has been growing at a rate of 5 percent per year for the last ten years. You might graph it this way:

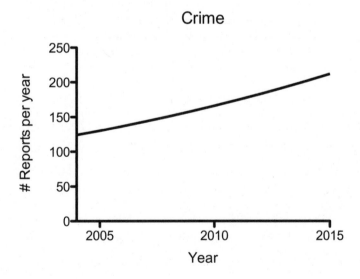

Nothing wrong with that. But suppose that you're selling home security systems and so you want to scare people into buying your product. Using all the same data, just create a discontinuity in your x-axis. This will distort the truth and deceive the eye marvelously:

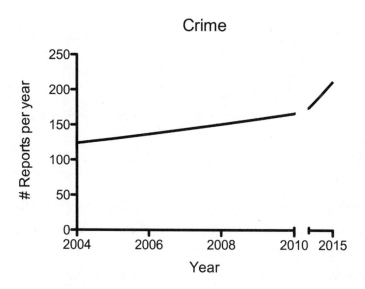

Here, the visual gives the impression that crime has increased dramatically. But you know better. The discontinuity in the x-axis crams five years' worth of numbers into the same amount of graphic real estate as was used for two years. No wonder there's an apparent increase. This is a fundamental flaw in graph making, but because most readers don't bother to look at the axes too closely, this one's easy to get away with.

And you don't have to limit your creativity to breaking the x-axis; you can get the effect by creating a discontinuity in the y-axis, and then hiding it by not breaking the line. While we're at it, we'll truncate the y-axis:

Crime

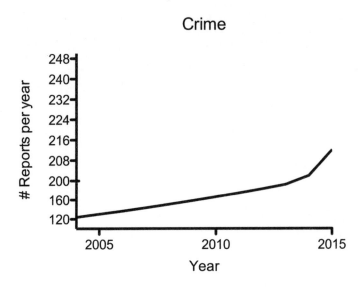

This is a bit mean. Most readers just look at that curve within the plot frame and won't notice that the tick marks on the vertical axis start out being forty reports between each, and then suddenly, at two hundred, indicate only eight reports between each. Are we having fun yet?

The honorable move is to use the first crime graph presented with the proper continuous axis. Now, to critically evaluate the statistics, you might ask if there are factors in the way the data were collected or presented that could be hiding an underlying truth.

One possibility is that the increases occur in only one particularly bad neighborhood and that, in fact, crime is *decreasing* everywhere else in the city. Maybe the police and the community have simply decided that a particular neighborhood had become unmanageable and so they stopped enforcing laws there. The city as a whole is safe—perhaps even safer than before—and one bad neighborhood is responsible for the increase.

Another possibility is that by amalgamating all the different sorts of complaints into the catchall bin of *crime*, we are overlooking a serious consideration. Perhaps *violent crime* has dropped to almost zero, and in its place, with so much time on their hands, the police are issuing hundreds more jaywalking tickets.

Perhaps the most obvious question to ask next, in your effort to understand what this statistic really means, is "What happened to the *total population* in this city during that time period?" If the population increased at any rate greater than 5 percent per year, the crime rate has actually gone down on a per-person basis. We could show this by plotting crimes committed per ten thousand people in the city:

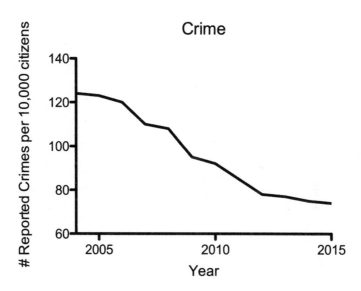

Choosing the Proper Scale and Axis

You've been hired by your local Realtor to graph the change in home prices in your community over the last decade. The prices have been steadily growing at a rate of 15 percent per year.

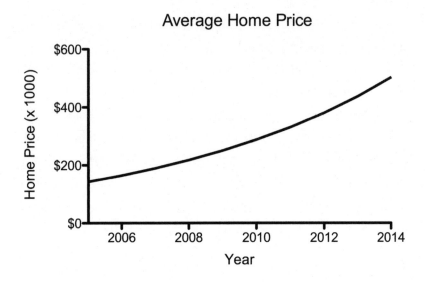

If you want to really alarm people, why not change the x-axis to include dates that you don't have data for? Adding extra dates to the x-axis artificially like this will increase the slope of the curve by compressing the viewable portion like this:

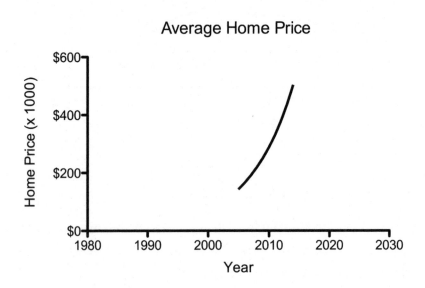

Notice how this graph tricks your eye (well, your brain) into drawing two false conclusions—first, that sometime around 1990 home prices must have been very low, and second, that by 2030 home prices will be so high that few people will be able to afford a home. Better buy one now!

Both of these graphs distort what's really going on, because they make a steady rate of growth appear, visually, to be an increasing rate of growth. On the first graph, the 15 percent growth seems twice as high on the y-axis in 2014 as it does in 2006. Many things change at a constant rate: salaries, prices, inflation, population of a species, and victims of diseases. When you have a situation of steady growth (or decline), the most accurate way to represent the data is on a logarithmic scale. The logarithmic scale allows equal percentage changes to be represented by equal distances on the y-axis. A constant annual rate of change then shows up as a straight line, as this:

The Dreaded Double Y-Axis

The graph maker can get away with all kinds of lies simply armed with the knowledge that most readers will not look at the graph very closely. This can move a great many people to believe all kinds of things that aren't so. Consider the following graph, showing the life expectancy of smokers versus nonsmokers at age twenty-five.

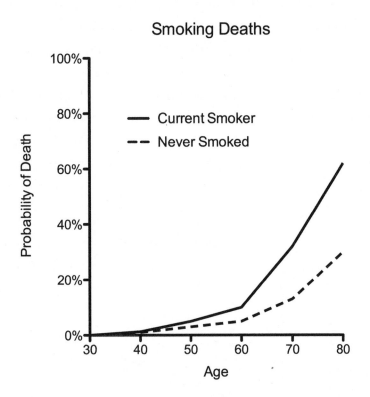

This makes clear two things: The dangers of smoking accumulate over time, and smokers are likely to die earlier than nonsmokers. The difference isn't big at age forty, but by age eighty the risk more than doubles, from under 30 percent to over 60 percent. This is a clean and accurate way to present the data. But suppose you're a young fourteen-year-old smoker who wants to convince your parents that you should be allowed to smoke. This graph is clearly not going to help you. So you dig deep into your bag of tricks and use the double y-axis, adding a y-axis to the right-hand side of the graph frame, with a different scaling factor that applies only to the non-smokers. Once you do that, your graph looks like this:

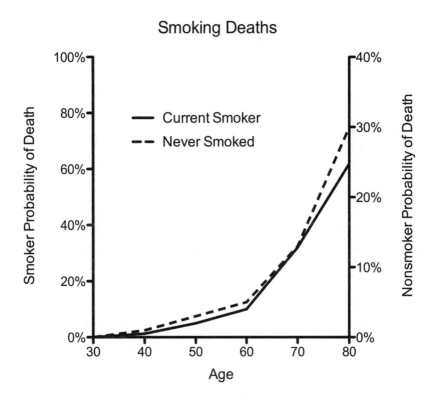

From this, it looks like you're just as likely to die from smoking as from not smoking. Smoking won't harm you—old age will! The trouble with double y-axis graphs is that you can always scale the second axis any way that you choose.

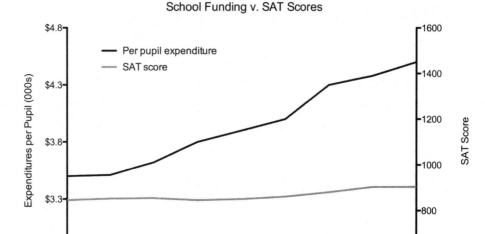

From the graph, it looks as though increasing the money spent per student (black line) doesn't do anything to increase their SAT scores (gray line). The story that some anti–government spending politicos could tell about this is one of wasted taxpayer funds. But you now understand that the choice of scale for the second (right-hand) y-axis is arbitrary. If you were a school administrator, you might simply take the exact same data, change the scale of the right-hand axis, and voilà—increasing spending delivers a better education, as evidenced by the increase in SAT scores!

Forbes magazine, a venerable and typically reliable news source, ran a graph very much like this one to show the relation between expenditures per public school student and those students' scores on the SAT, a widely used standardized test for college admission in the United States.

From the graph, it looks as though increasing the money spent per student (black line) doesn't do anything to increase their SAT scores (gray line). The story that some anti–government spending politicos could tell about this is one of wasted taxpayer funds. But you now understand that the choice of scale for the second (right-hand) y-axis is arbitrary. If you were a school administrator, you might simply take the exact same data, change the scale of the right-hand axis, and voilà—increasing spending delivers a better education, as evidenced by the increase in SAT scores!

School Funding v. SAT Scores

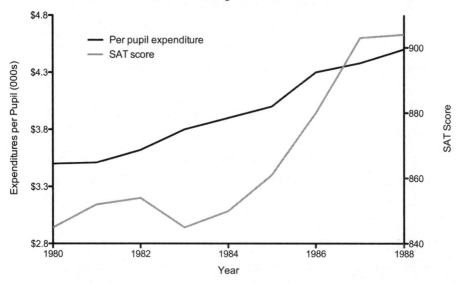

This graph obviously tells a very different story. Which one is true? You'd need to have a measure of how the one variable changes as a function of the other, a statistic known as a correlation. Correlations range from –1 to 1. A correlation of 0 means that one variable is not related to the other at all. A correlation of -1 means that as one variable goes up, the other goes down, in precise synchrony. A correlation of 1 means that as one variable goes up, the other does too, also in precise synchrony. The first graph appears to be illustrating a correlation of 0, the second graph appears to be representing one that is close to 1. The actual correlation for this dataset is .91, a very strong correlation. Spending more on students is, at least in this dataset, associated with better SAT scores.

The correlation also provides a good estimate of how much of the result can be explained by the variables you're looking at. The correlation of .91 tells us we can explain 91 percent of students' SAT scores by looking at the amount of school expenditures per student. That is, it tells us to what extent expenditures explain the diversity in SAT scores.

A controversy about the double y-axis graph erupted in the fall of 2015 during a U.S. congressional committee meeting. Rep. Jason Chaffetz presented a graph that plotted two services provided by the organization Planned Parenthood: abortions, and cancer screening and prevention:

PLANNED PARENTHOOD FEDERATION OF AMERICA:
ABORTIONS UP — LIFE-SAVING PROCEDURES DOWN

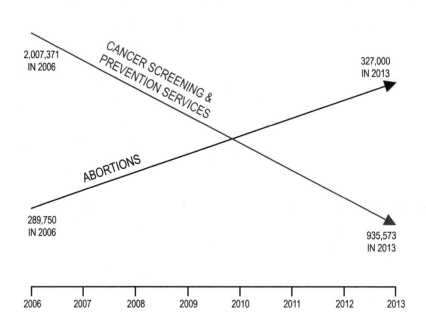

2,007,371 IN 2006

CANCER SCREENING & PREVENTION SERVICES

327,000 IN 2013

ABORTIONS

289,750 IN 2006

935,573 IN 2013

2006   2007   2008   2009   2010   2011   2012   2013

The congressman was attempting to make a political point, that over a seven-year period, Planned Parenthood has increased the number of abortions it performed (something he opposes) and decreased the number of cancer screening and prevention procedures. Planned Parenthood doesn't deny this, but this distorted graph makes it seem that the number of abortion procedures exceeded those for cancer. Maybe the graph maker was feeling a bit guilty and so included the actual numbers next to the data points. Let's accept her bread crumbs and look closely. The number of abortions in 2013, the most recent year given, is 327,000. The number of cancer services was nearly three times that, at 935,573. (By the way, it's a bit suspicious that the abortion numbers are such tidy, round numbers while the cancer numbers are so precise.) This is a particularly sinister example: an implied double y-axis graph with no axes on either side!

Drawn properly, the graph would look like this:

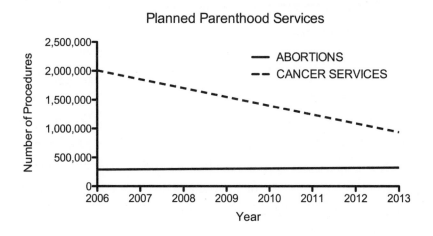

Here, we see that abortions increased modestly, compared to the reduction in cancer services.

There is another thing suspicious about the original graph: Such smooth lines are rarely found in data. It seems more likely that the graph maker simply took numbers for two particular years, 2006 and 2013, and compared them, drawing a smooth connecting line between them. Perhaps these particular years were chosen intentionally to emphasize differences. Perhaps there were great fluctuations in the intervening years of 2007–2012; we don't know. The smooth lines give the impression of a perfectly linear (straight line) function, which is very unlikely.

Graphs such as this do not always tell the story that people think they do. Is there something that could account for these data, apart from a narrative that Planned Parenthood is on a mission to perform as many abortions as it can (and to let people die of cancer at the same time)? Look at the second graph. In 2006, Planned Parenthood performed 2,007,371 cancer services, and 289,750 abortions, nearly seven times as many cancer services as abortions. By 2013, this gap had narrowed, but the number of cancer services was still nearly three times the number of abortions.

Cecile Richards, the president of Planned Parenthood, had an explanation for this narrowing gap. Changing medical guidelines for some anti-cancer services, like Pap smears, reduced the number of people for whom screening was recommended. Other changes, such as social attitudes about abortion, changing ages of the population, and increased access to health care alternatives, all influence these numbers, and so the data presented do not prove that Planned Parenthood has a pro-abortion agenda. It might—these data are just not the proof.

HIJINKS WITH HOW NUMBERS ARE REPORTED

You're trying to decide whether to buy stock in a new soft drink and you come across this graph of the company's sales figures in their annual report:

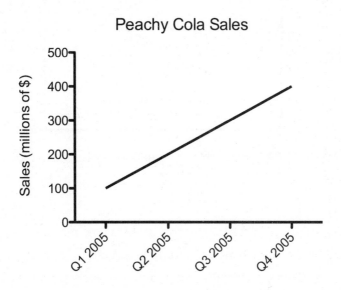

This looks promising—Peachy Cola is steadily increasing its sales. So far, so good. But a little bit of world knowledge can be applied here to good effect. The soft-drink market is very competitive. Peachy Cola's sales are increasing, but maybe not as quickly as a competitor's. As a potential investor, what you really want to see is how Peachy's sales compare to those of other companies, or to see their sales as a function of market share—Peachy's sales could go up only slightly while the market is growing enormously, and competitors are benefiting more than Peachy is. And, as this example of a useful double y-axis graph demonstrates, this may not bode well for their future:

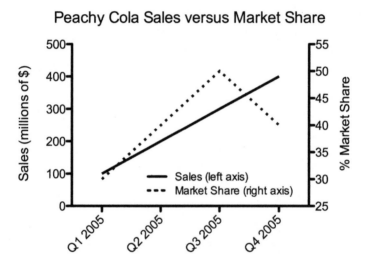

Although unscrupulous graph makers can monkey with the scaling of the right-hand axis to make the graph appear to show anything they want, this kind of double-y-axis graph isn't scandalous because the two y-axes are representing different things, quantities that *couldn't* share an axis. This was not the case with the Planned Parenthood graph on page 40, which was reporting the same quantity on the two different axes, the number of performed procedures. That graph was distorted by ensuring that the two axes, although they measure the same thing, were scaled differently in order to manipulate perception.

It would also be useful to see Peachy's profits: Through manufacturing and distribution efficiencies, it may well be that they're making more money on a lower sales volume. Just because someone quotes you a statistic or shows you a graph, it doesn't mean it's relevant to the point they're trying to make. It's the job of all of us to make sure we get the information that matters, and to ignore the information that doesn't.

Let's say that you work in the public-affairs office for a company that manufactures some kind of device—frabezoids. For the last several years, the public's appetite for frabezoids has been high, and sales have increased. The company expanded by building new facilities, hiring new employees, and giving everyone a raise. Your boss comes into your cubicle with a somber-looking expression and explains that the newest sales results are in, and frabezoid sales have dropped 12 percent from the previous quarter. Your company's president is about to hold a big press conference to talk about the future of the company. As is his custom, he'll display a large graph on the stage behind him showing how frabezoids are doing. If word gets out about the lower sales figures, the public may think that frabezoids are no longer desirable things to have, which could then lead to an even further decline in sales.

What do you do? If you graph the sales figures honestly for the past four years, your graph would look like this:

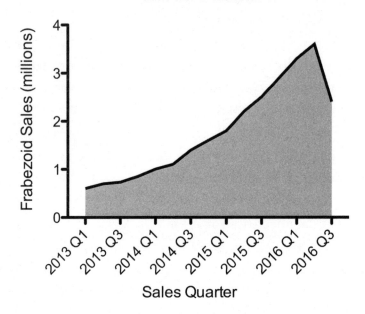

That downward trend in the curve is the problem. If only there were a way to make that curve go up.

Well, there is! The cumulative sales graph. Instead of graphing sales per quarter, graph the cumulative sales per quarter—that is, the total sales to date.

As long as you sold only one frabezoid, your cumulative graph will increase, like this one here:

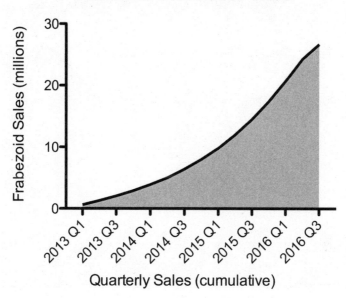

If you look carefully, you can still see a vestige of the poor sales for last quarter: Although the line is still going up for the most recent quarter, it's going up less steeply. That's your clue that sales have dropped. But our brains aren't very good at detecting rates of change such as these (what's called the first derivative in calculus, a fancy name for the slope of the line). So on casual examination, it seems the company continues to do fabulously well, and you've made a whole lot of consumers believe that frabezoids are still the hottest thing to have.

This is exactly what Tim Cook, CEO of Apple, did recently in a presentation on iPhone sales.

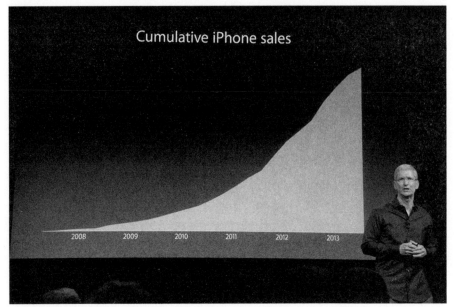

© 2013 The Verge, Vox Media Inc. (live.theverge.com/apple-iphone-5s-liveblog/)

Plotting Things That Are Unrelated

There are so many things going on in the world that some coincidences are bound to happen. The number of green trucks on the road may be increasing at the same time as your salary; when you were a kid, the number of shows on television may have increased with your height. But that doesn't mean that one is causing the other. When two things are related, whether or not one causes the other, statisticians call it a correlation.

The famous adage is that "correlation does not imply causation." In formal logic there are two formulations of this rule:

1) *Post hoc, ergo propter hoc* (after this, therefore because of this). This is a logical fallacy that arises from thinking that just because

one thing (Y) occurs after another (X), that X *caused* Y. People typically brush their teeth before going off to work in the morning. But brushing their teeth doesn't *cause* them to go to work. In this case, it is even possibly the reverse.

2) *Cum hoc, ergo propter hoc* (with this, therefore because of this). This is a logical fallacy that arises from thinking that just because two things co-occur, one must have caused the other. To drive home the point, Harvard Law student Tyler Vigen has written a book and a website that feature spurious co-occurrences—correlations—such as this one:

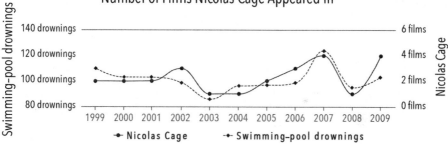

There are four ways to interpret this: (1) drownings cause the release of new Nicolas Cage films; (2) the release of Nicolas Cage films causes drownings; (3) a third (as yet unidentified) factor causes both; or (4) they are simply unrelated and the correlation is a coincidence. If we don't separate correlation from causation, we can claim that Vigen's graph "proves" that Nic Cage was helping to prevent pool drownings, and furthermore, our best bet is to encourage him to make fewer movies so that he can ply his lifesaving skills as he apparently did so effectively in 2003 and 2008.

In some cases, there is no actual connection between items that are correlated—their correlation is simply coincidence. In other cases, one can find a causal link between correlated items, or at least spin a reasonable story that can spur the acquisition of new data.

We can rule out explanation one, because it takes time to produce and release a movie, so a spike in drownings cannot cause a spike in Nic Cage movies in the same year. What about number two? Perhaps people become so wrapped up in the drama of Cage's films that they lose focus and drown as a consequence. It may be that the same cinematic absorption also increases rates of automobile accidents and injuries from heavy machinery. We don't know until we analyze more data, because those are not reported here.

What about a third factor that caused both? We might guess that economic trends are driving both: A better economy leads to more investment in leisure activities—more films being made, more people going on vacation and swimming. If this is true, then neither of the two things depicted on the graph—Nic Cage films and drownings—caused the other. Instead, a third factor, the economy, led to changes in both. Statisticians call this the *third factor x* explanation of correlations, and there are many cases of these.

More likely, these two are simply unrelated. If we look long enough, and hard enough, we're sure to find that two unrelated things vary with each other.

Ice-cream sales increase as the number of people who wear short pants increases. Neither causes the other; the third factor *x* that causes both is the warmer temperatures of summer. The number of television shows aired in a year while you were a child may have correlated with increases in your height, but what was no doubt driving both was the passage of time during an era when (a) TV was expanding its market and (b) you were growing.

How do you know when a correlation indicates causation? One way is to conduct a controlled experiment. Another is to apply logic. But be careful—it's easy to get bogged down in semantics. Did the rain outside *cause* people to wear raincoats or was it their desire to avoid getting wet, a consequence of the rain, that caused it?

This idea was cleverly rendered by Randall Munroe in his Internet cartoon *xkcd*. Two stick figures, apparently college students, are talking. One says that he used to think correlation implied causation. Then he took a statistics class, and now he doesn't think that anymore. The other student says, "Sounds like the class helped." The first student replies, "Well, maybe."

Deceptive Illustrations

Infographics are often used by lying weasels to shape public opinion, and they rely on the fact that most people won't study what they've done too carefully. Consider this graphic that might be used to scare you into thinking that runaway inflation is eating up your hard-earned money:

That's a frightening image. But look closely. The scissors are cutting the bill not at 4.2 percent of its size, but at about 42 percent. When your visual system is pitted against your logical system, the visual system usually wins, unless you work extra diligently to overcome this visual bias. The accurate infographic would look like this but would have much less emotional impact:

Interpreting and Framing

Often a statistic will be properly created and reported, but someone—a journalist, an advocate, any non-statistician—will misreport it, either because they've misunderstood it or because they didn't realize that a small change in wording can change the meaning.

Often those who want to use statistics do not have statisticians on their staffs, and so they seek the answers to their questions from people who lack proper training. Corporations, government offices, nonprofits, and mom-and-pop grocery stores all benefit from statistics about such items as sales, customers, trends, and supply chain. Incompetence can enter at any stage, in experimental design, data collection, analysis, or interpretation.

Sometimes the statistic being reported isn't the relevant one. If you're trying to convince stockholders that your company is doing well, you might publish statistics on your annual sales, and show steadily rising numbers. But if the market for your product is expanding, sales increases would be expected. What your investors and analysts probably want to know is whether your market share has changed. If your market share is decreasing because competitors are swooping in and taking away your customers, how can you make your report look attractive? Simply fail to report the relevant statistic of market share, and instead report the sales figures. Sales are going up! Everything is fine!

The financial profiles shown on people's mortgage applications twenty-five years ago would probably not be much help in building a model for risk today. Any model of consumer behavior on a website may become out of date very quickly. Statistics on the integrity of concrete used for overpasses may not be relevant for concrete on bridges (where humidity and other factors may have caused divergence, even if both civic projects used the same concrete to begin with).

You've probably heard some variant of the claim that "four out of five dentists recommend Colgate toothpaste." That's true. What the ad agency behind these decades-old ads wants you to think is that the dentists prefer Colgate above and beyond other brands. But that's not true. The Advertising Standards Authority in the United Kingdom investigated this claim and ruled it an unfair practice because the survey that was conducted allowed dentists to recommend more than one toothpaste. In fact, Colgate's biggest competitor was named nearly as often as Colgate (a detail you won't see in Colgate's ads).

Framing came up in the section on averages and implicitly in the discussion of graphs. Manipulating the framing of any message furnishes an endless number of ways people can make you believe something that isn't so if you don't stop to think about what they're saying. The cable network C-SPAN advertises that it is "available" in 100 million homes. That doesn't mean that 100 million people are watching C-SPAN. It doesn't mean that even one person is watching it.

Framing manipulations can influence public policy. A survey of recycling yield on various streets in metropolitan Los Angeles shows that one street in particular recycles 2.2 times as much as any other street. Before the city council gives the residents of this street an award for their green city efforts, let's ask what might give rise to such a number. One possibility is that this street has more than twice as many residents as other streets—perhaps because it is longer, perhaps because there are a lot of apartment buildings on it. Measuring recycling at the level of the street is not the relevant statistic unless all streets are otherwise identical. A better statistic would be either the living unit (where you measure the recycling output of each family) or even better, because larger families probably consume more than smaller families, the individual. That is, we want to adjust the amount of recycling materials collected to take into account the number of people on the street. That is the *true frame* for the statistic.

The *Los Angeles Times* reported in 2014 about water use in the city of Rancho Santa Fe in drought-plagued California. "On a daily per capita basis, households in this area lapped up an average of nearly five times the water used by coastal Southern California homes in September, earning them the dubious distinction of being the state's biggest residential water hogs." "Households" is not the relevant frame for this statistic, and the *LA Times* was correct to report per capita—individuals; perhaps the residents of Rancho Santa Fe have larger families, meaning more showers, dishes, and flushing commodes. Another frame would look at water use per acre. Rancho Santa Fe homes tend to have larger lots. Perhaps it is desirable for fire prevention and other reasons to keep land planted with verdant vegetation, and the large lots in Rancho Santa Fe don't use more water on a per acre basis than land anywhere else.

In fact, there's a hint of this in a *New York Times* article on the issue: "State water officials warned against comparing per capita water use between districts; they said they expected use to be highest in wealthy communities with large properties."

The problem with the newspaper articles is that they frame the data to make it look as though Rancho Santa Fe residents are using more than their share of water, but the data they provide—as in the case of the Los Angeles recycling example above—don't actually show that.

Calculating proportions rather than actual numbers often helps to provide the true frame. Suppose you are northwest regional sales manager for a company that sells flux capacitors. Your sales have improved greatly, but are still no match for your nemesis in the company, Jack from the southwest. It's hardly fair—his territory is not only geographically larger but covers a much larger population. Bonuses in your company depend on you showing the higher-ups that you have the mettle to go out and get sales.

There is a legitimate way to present your case: Report your sales as a function of the area or population of the territory you serve. In other words, instead of graphing total number of flux capacitors sold, look at total number per person in the region, or per square mile. In both, you may well come out ahead.

News reports showed that 2014 was one of the deadliest years for plane crashes: 22 accidents resulted in 992 fatalities. But flying is actually safer now than it has ever been. Because there are so many more flights today than ever before, the 992 fatalities represent a dramatic decline in the number of deaths per million passengers (or per million miles flown). On any single flight on a major airline, the chances are about 1 in 5 million that you'll be killed, making it more likely that you'll be killed doing just about anything else—walking across the street, eating food (death by choking or unintentional poisoning is about 1,000 times more likely). The baseline for comparison is very important here. These statistics are spread out over a year—a year of airline travel, a year of eating and then either choking or being poisoned. We could change the baseline and look at each hour of the activities, and this would change the statistic.

Differences That Don't Make a Difference

Statistics are often used when we seek to understand whether there is a difference between two treatments: two different fertilizers in a field, two different pain medications, two different styles of teaching, two different groups of salaries (e.g., men versus women doing the same jobs). There are many ways that two treatments can differ. There can be actual differences between them; there can be confounding factors in your sample that have nothing to do with the actual treatments; there can be errors in your measurement; or there can be random variation—little chance differences that turn up, sometimes on one side of the equation, sometimes on the other, depending on when you're looking. The researcher's goal is to find stable, replicable differences, and we try to distinguish those from experimental error.

Be wary, though, of the way news media use the word "significant," because to statisticians it doesn't mean "noteworthy." In statistics, the word "significant" means that the results passed mathematical tests such as t-tests, chi-square tests, regression, and principal components analysis (there are hundreds). Statistical significance tests quantify how easily pure chance can explain the results. With a very large number of observations, even small differences that are trivial in magnitude can be beyond what our models of change and randomness can explain. These tests don't know what's noteworthy and what's not—that's a human judgment.

The more observations you have in the two groups, the more likely that you will find a difference between them. Suppose I test the annual maintenance costs of two automobiles, a Ford and a Toyota, by looking at the repair records for ten of each car. Let's say, hypothetically, the mean cost of operating the Ford is eight cents more per year. This will probably fail to meet statistical significance, and clearly a cost difference of eight cents a year is not going to be the deciding factor in which car to buy—it's just too small an amount to be concerned about. But if I look at the repair records for 500,000 vehicles, that eight-cent difference will be statistically significant. But it's a difference that doesn't matter in any real-world, practical sense. Similarly, a new headache medication may be statistically faster at curing your headache, but if it's only 2.5 seconds faster, who cares?

Interpolation and Extrapolation

You go out in your garden and see a dandelion that's four inches high on Tuesday. You look again on Thursday and it's six inches high. How high was it on Wednesday? We don't know for sure because we didn't measure it Wednesday (Wednesday's the day you got stuck in traffic on the way home from the nursery, where you bought some weed killer). But you can guess: The dandelion was probably five inches high on Wednesday. This is interpolation. Interpolation takes two data points and estimates the value that would have occurred between them if you had taken a measurement there.

How high will the dandelion be after six months? If it's growing 1 inch per day, you might say that it will grow 180 inches more in six months (roughly 180 days), for a total of 186 inches, or fifteen and a half feet high. You're using extrapolation. But have you ever seen a dandelion that tall? Probably not. They collapse under their own weight, or die of other natural causes, or get trampled, or the weed killer might get them. Interpolation isn't a perfect technique, but if the two observations you're considering are very close together, interpolation usually provides a good estimate. Extrapolation, however, is riskier, because you're making estimates outside the range of your observations.

The amount of time it takes a cup of coffee to cool to room temperature is governed by Newton's law of cooling (and is affected by other factors such as the barometric pressure and the composition of the cup). If your coffee started out at 145 degrees Fahrenheit (F), you'd observe the temperature decreasing over time like this:

Elapsed Time (mins)	Temp °F
0	145
1	140
2	135
3	130

Your coffee loses five degrees every minute. If you interpolated between two observations—say you want to know what the temperature would have been at the halfway point between measurements—your interpolation is going to be quite accurate. But if you extrapolate from the pattern, you are likely to come up with an absurd answer, such as that the coffee will reach freezing after thirty minutes.

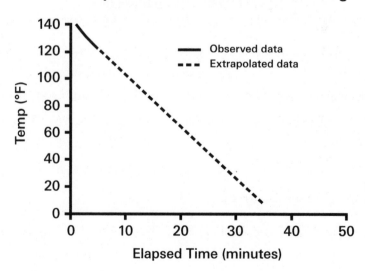

Temperature of Coffee Left Standing

The extrapolation fails to take into account a physical limit: The coffee can't get cooler than room temperature. It also fails to take into account that the rate at which the coffee cools slows down the closer it gets to room temperature. The rest of the cooling function looks like this:

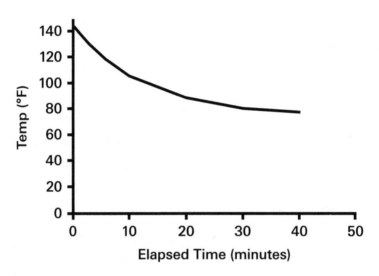

Temperature of Coffee Left Standing

Note that the steepness of the curve in the first ten minutes doesn't continue—it flattens out. This underscores the importance of two things when you're extrapolating: having a large number of observations that span a wide range, and having some knowledge of the underlying process.

Precision Versus Accuracy

When faced with the precision of numbers, we tend to believe that they are also *accurate*, but this is not the same thing. If I say "a lot of people are buying electric cars these days," you assume that I'm making a guess. If I say that "16.39 percent of new car sales are electric vehicles," you assume that I know what I'm talking about. But you'd be confusing precision for accuracy. I may have made it up. I may have sampled only a small number of people near an electric-car dealership.

Recall the *Time* magazine headline I mentioned earlier, which said that more people have cell phones than have toilets. This isn't implausible, but it is a distortion because that's *not* what the U.N. study found at all. The U.N. reported that more people had *access* to cell phones than to toilets, which is, as we know, a different thing. One cell phone might be shared among dozens of people. The lack of sanitation is still distressing, but the headline makes it sound like if you were to count, you'd find there are more cell phones in the world than there are toilets, and that is not supported by the data.

Access is one of those words that should raise red flags when you encounter them in statistics. People having access to health care might simply mean they live near a medical facility, not that the facility would admit them or that they could pay for it. As you learned above, C-SPAN is available in 100 million homes, but that doesn't mean that 100 million people are watching it. I could claim that 90 percent of the world's population has "access" to *A Field Guide to Lies* by showing that 90 percent of the population is within twenty-five miles of an Internet connection, rail line, road, landing strip, port, or dogsled route.

Comparing Apples and Oranges

One way to lie with statistics is to compare things—datasets, populations, types of products—that are different from one another, and pretend that they're not. As the old idiom says, you can't compare apples with oranges.

Using dubious methods, you could claim that it is safer to be in the military during an active conflict (such as the present war in Afghanistan) than to be stateside in the comfort of your own home. Start with the 3,482 active-duty U.S. military personnel who died in 2010. Out of a total of 1,431,000 people in the military, this gives a rate of 2.4 deaths per 1,000. Across the United States, the death rate in 2010 was 8.2 deaths per 1,000. In other words, it is more than three times safer to be in the military, in a war zone, than to live in the United States.

What's going on here? The two samples are not similar, and so shouldn't be compared directly. Active military personnel tend to be young and in good health; they are served a nutritious diet and have good health care. The general population of the United States includes the elderly, people who are sick, gang members, crackheads, motorcycle daredevils, players of mumblety-peg, and many people who have neither a nutritious diet nor good health care; their mortality rate would be high wherever they are. And active military personnel are not all stationed in a war zone—some are stationed in very safe bases in the United States, are sitting behind desks in the Pentagon, or are stationed in recruiting stations in suburban strip malls.

U.S. News & World Report published an article comparing the proportion of Democrats and Republicans in the country going back to the 1930s. The problem is that sampling methods have changed over the years. In the 1930s and '40s, sampling was typically done by in-person interviews and mail lists generated by telephone directories; by the 1970s sampling was predominantly just by telephone. Sampling in the early part of the twentieth century skewed toward those who tended to have landlines: wealthier people, who, at least at that time, tended to vote Republican. By the 2000s, cell phones were being sampled, which skewed toward the young, who tended to vote Democratic. We can't really know if the proportion of Democrats to Republicans has changed since the 1930s because the samples are incompatible. We think we're studying one thing but we're studying another.

A similar problem occurs when reporting a decline in the death rate due to motorcycle accidents now versus three decades ago. The more recent figures might include more three-wheel motorcycles compared to predominantly two-wheeled ones last century; it might compare an era when helmets were not required by law to now, where they are in most states.

Be on the lookout for changing samples before drawing conclusions! *U.S. News & World Report* (yes, them again) wrote of an increase in the number of doctors over a twelve-year period, accompanied by a significant drop in average salary. What is the takeaway message? You might conclude that now is not a good time to enter the medical profession because there is a glut of doctors, and that supply exceeding demand has lowered every doctor's salary. This might be true, but there is no evidence in the claim to support this.

An equally plausible argument is that over the twelve-year period, increased specialization and technology growth created more opportunities for doctors and so there were more available positions, accounting for the increase in the total number of doctors. What about the salary decline? Perhaps many older doctors retired, and were replaced by younger ones, who earn a smaller salary just out of medical school. There is no evidence presented either way. An important part of statistical literacy is recognizing that some statistics, as presented, simply cannot be interpreted.

Sometimes, this apples-and-oranges comparison results from

inconsistent subsamples—ignoring a detail that you didn't realize was important. For example, when sampling corn from a field that received a new fertilizer, you might not notice that some ears of corn get more sun and some get more water. Or when studying how traffic patterns affect street repaving, you might not realize that certain streets have more water runoff than others, influencing the need for asphalt repairs.

Amalgamating is putting things that are different (heterogeneous) into the same bin or category—a form of apples and oranges. If you're looking at the number of defective sprockets produced by a factory, you might combine two completely different kinds in order to make the numbers come out more favorably for your particular interests.

Take an example from public policy. You might want to survey the sexual behavior of preteens and teens. How you amalgamate (or bin) the data can have a large effect on how people perceive your data. If your agenda is to raise money for educational and counseling centers, what better way to do so than to release a statistic such as "70 percent of schoolchildren ages ten to eighteen are sexually active." We're not surprised that seventeen- and eighteen-year-olds are, but ten-year-olds! That will surely cause grandparents to reach for the smelling salts and start writing checks. But obviously, a single category of ten-year-olds to eighteen-year-olds lumps together individuals who are likely to be sexually active with those who are not. More helpful would be separate bins that put together individuals of similar age and likely similar experiences: ten to eleven, twelve to thirteen, fourteen to fifteen, sixteen to eighteen, for example.

But that's not the only problem. What do they mean by "sexually active"? What question was actually asked of the schoolchildren?

Or were the schoolchildren even asked? Perhaps it was their parents who were asked. All kinds of biases can enter into such a number. "Sexually active" is open to interpretation. Responses will vary widely depending on how it is defined. And of course respondants may not tell the truth (reporting bias).

As another example, you might want to talk about unemployment as a general problem, but this risks combining people of very different backgrounds and contributing factors. Some are disabled and can't work; some are fired with good cause because they were caught stealing or drunk on the job; some want to work but lack the training; some are in jail; some no longer want to work because they've gone back to school, joined a monastery, or are living off family money. When statistics are used to influence public policy, or to raise donations for a cause, or to make headlines, often the nuances are left out. And they can make all the difference.

These nuances often tell a story themselves about patterns in the data. People don't become unemployed for the same reasons. The likelihood that an alcoholic or a thief will become unemployed may be four times that of someone who is not. These patterns carry information that is lost in amalgamation. Allowing these factors to become part of the data can help you to see who is unemployed and why—it could lead to better training programs for people who need it, or more Alcoholics Anonymous centers in a town that is underserved by them.

If the people and agencies who track behavior use different definitions for things, or different procedures for measuring them, the data that go into the statistic can be very dissimilar, or heterogeneous. If you're trying to pin down the number of couples who live together but are not married, you might rely on data that have already been collected by various county and state agencies. But varying definitions can yield a categorization problem: What constitutes living together? Is it determined by how many nights a week they are together? By where their possessions are, where they get mail? Some jurisdictions recognize same-sex couples and some don't. If you take the data from different places using different schemes, the final statistic carries very little meaning. If the recording, collection, and measurement practices vary widely across collection points, the statistic that results may not mean what you think it means.

A recent report found that the youth unemployment rate in Spain was an astonishing 60 percent. The report amalgamated into the same category people who normally would appear in separate categories: Students who were not seeking work were counted as unemployed, alongside workers who had just been laid off and workers who were seeking jobs.

In the United States, there are *six* different indexes (numbered U1 through U6) to track unemployment (as measured by the Bureau of Labor Statistics), and they reflect different interpretations of what "unemployed" actually means. It can include people looking for a job, people who are in school but not looking, people who are seeking full-time assignments in a company where they work only part-time, and so on.

USA Today reported in July 2015 that the unemployment rate dropped to 5.3 percent, "its lowest level since April 2008." More comprehensive sources, including the AP, *Forbes*, and the *New York Times*, reported the reason for the apparent drop: Many people who were out of work gave up looking and so technically had left the workforce.

Amalgamating isn't always wrong. You might choose to combine the test scores of boys and girls in a school, especially if there is no evidence that their scores differ—in fact, it's a good idea to, in order to increase your sample size (which provides you with a more stable estimate of what you're studying). Overly broad definitions of a category (as with the sexual-activity survey mentioned earlier) or inconsistent definitions (as with the couples-living-together statistic) present problems for interpretation. When performed properly, amalgamating helps us come up with a valid analysis of data.

Suppose that you work for the state of Utah and a large national manufacturer of baby clothes is thinking about moving to your state. You're thinking that if you can show that Utah has a lot of births, you're in a better position to attract the company, so you go to the Census.gov website, and graph the results for number of births by state:

Births: United States, 2013

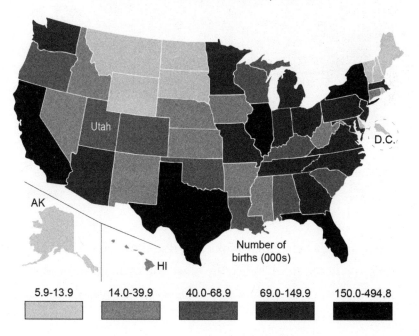

Number of births (000s)

| 5.9-13.9 | 14.0-39.9 | 40.0-68.9 | 69.0-149.9 | 150.0-494.8 |

Utah looks better than Alaska, D.C., Montana, Wyoming, the Dakotas, and the small states of the Northeast. But it is hardly a booming baby state compared to California, Texas, Florida, and New York. But wait, this map you've made shows the *raw number* of births and so will be weighted heavily toward states with larger populations. Instead, you could graph the birth *rate* per thousand people in the population:

Crude Birth Rate: United States, 2013

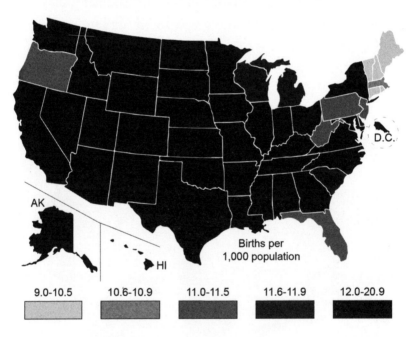

That doesn't help. Utah looks just like most of the rest of the country. What to do? Change the bins! You can play around with which range of values go into each category, those five gray-to-black bars at the bottom. By making sure that Utah's rate is in a category all by itself, you can make it stand out from the rest of the country.

Crude Birth Rate: United States, 2013

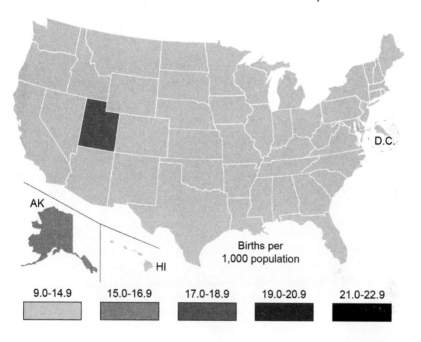

Of course, this only works because Utah does in fact have the highest birth rate in the country—not by much, but it is still the highest. By choosing a bin that puts it all by itself in a color category, you've made it stand out. If you were trying to make a case for one of the other states, you'd have to resort to other kinds of flimflam, such as graphing the number of births per square mile, or per Walmart store, as a function of disposable income. Play around long enough and you might find a metric to make a case for any of the fifty states.

What is the *right* way, the non-lying way to present such a graph? This is a matter of judgment, but one relatively neutral way would be to bin the data so that 20 percent of the states are contained in each of the five bins, that is, an equal number of states per color category:

Crude Birth Rate: United States, 2013

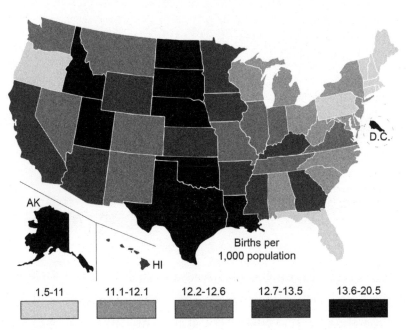

| 1.5-11 | 11.1-12.1 | 12.2-12.6 | 12.7-13.5 | 13.6-20.5 |

Another would be to make the bins equal in size:

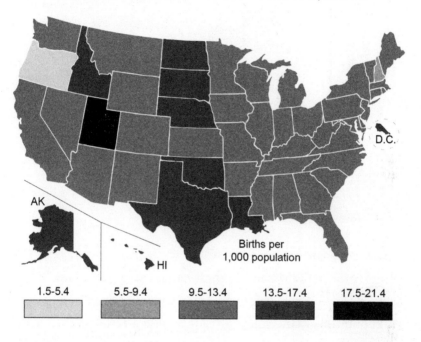

Crude Birth Rate: United States, 2013

Births per
1,000 population

| 1.5-5.4 | 5.5-9.4 | 9.5-13.4 | 13.5-17.4 | 17.5-21.4 |

This kind of statistical chicanery—using unequal bin widths in all but the last of these maps—often shows up in histograms, where the bins are typically identified by their midpoint and you have to infer the range yourself. Here are the batting averages for the 2015 season for the Top 50 qualifying Major League Baseball players (National and American Leagues):

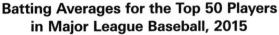

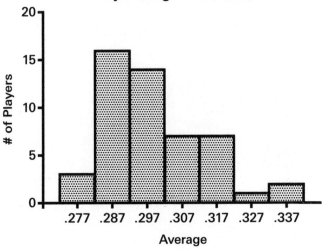

Batting Averages for the Top 50 Players in Major League Baseball, 2015

Now, suppose that you're the player whose batting average is .330, putting you in the second highest category. It's time for bonus checks and you don't want to give management any reason to deny you a bonus this year—you've already bought a Tesla. So change the bin widths, amalgamating your results with the two players who were batting .337, and now you're in with the very best players. While you're at it, close up the ensuing gap (there are no longer any batters in the .327-centered bin), creating a discontinuity in the x-axis that probably few will notice:

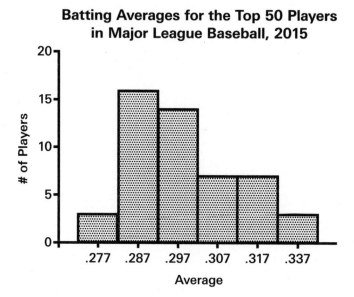

Batting Averages for the Top 50 Players in Major League Baseball, 2015

Specious Subdividing

The opposite of amalgamating is subdividing, and this can cause people to believe all kinds of things that aren't so. To claim that x is a leading cause of y, I simply need to subdivide other causes into smaller and smaller categories.

Suppose you work for a manufacturer of air purifiers, and you're on a campaign to prove that respiratory disease is the leading cause of death in the United States, overwhelming other causes like heart disease and cancer. As of today, the actual leading cause of death in the United States is heart disease. The U.S. Centers for Disease Control report that these were the top three causes of death in 2013:

Heart disease: 611,105

Cancer: 584,881

Chronic lower respiratory diseases: 149,205

Now, setting aside the pesky detail that home air purifiers may not form a significant line of defense against chronic respiratory disease, these numbers don't make a compelling case for your company. Sure, you'd like to save more than 100,000 lives annually, but to say that you're fighting the *third* largest cause of death doesn't make for a very impressive ad campaign. But wait! Heart disease isn't one thing, it's several:

Acute rheumatic fever and chronic rheumatic heart disease:
   3,260

Hypertensive heart disease: 37,144

Acute myocardial infarction: 116,793

Heart failure: 65,120

And so on. Next, break up the cancers into small subtypes. By failing to amalgamate, and creating these fine subdivisions, you've done it! Chronic lower respiratory disease becomes the number one killer. You've just earned yourself a bonus. Some food companies have used this subdivide strategy to hide the amounts of fats and sugars contained in their product.

How Numbers Are Collected

Just because there's a number on it, it doesn't mean that the number was arrived at properly. Remember, as the opening to this part of the book states, *people* gather statistics. People choose what to count, how to go about counting. There are a host of errors and biases that can enter into the collection process, and these can lead millions of people to draw the wrong conclusions. Although most of us won't ever participate in the collection process, thinking about it, critically, is easy to learn and within the reach of all of us.

Statistics are obtained in a variety of ways: by looking at records (e.g., birth and death records from a government agency, hospital, or church), by conducting surveys or polls, by observation (e.g., counting the number of electric cars that pass the corner of Main and Third Street), or by inference (if sales of diapers are going up, the birth rate is probably going up). Biases, inaccuracies, and honest mistakes can enter at any stage. Part of evaluating claims includes asking the questions "Can we really know that?" and "How do they know that?"

Sampling

Astrogeologists sample specimens of moon rock; they don't test the entire moon. Researchers don't talk to every single voter to find out which candidate is ahead, or tally every person who enters an emergency room to see how long they had to wait to be seen. To do so would be impractical or too costly. Instead, they use samples to estimate the true number. When samples are properly taken, these estimates can be very, very accurate. In public-opinion polls, an estimate of how the entire country feels about an issue (about 234 million adults over the age of twenty-one) can be obtained by interviewing only 1,067 individuals. Biopsies that sample less than one one-thousandth of an organ can be used for accurate cancer staging.

To be any good, a sample has to be representative. A sample is representative if every person or thing in the group you're studying has an equally likely chance of being chosen. If not, your sample is biased. If the cancer is only in part of an organ and you sample the wrong part, the cancer will go undiagnosed. If it's in a very small part and you take fifteen samples in that one spot, you may be led to conclude the entire organ is riddled with cancer when it's not.

We don't always know ahead of time, with biopsies or public-opinion polls, how much variability there will be. If everyone in a population were identical, we'd only need to sample one of them. If we have a bunch of genetically identical people, with identical personalities and life experience, we can find out anything we want to know about all of them by simply looking at one of them. But every group contains some heterogeneity, some differences across its members, and so we need to be careful about how we sample to ensure that we have accounted for all the differences that matter. (Not all differences matter.) For example, if we deprive a human of oxygen, we know that human will die. Humans don't differ along this dimension (although they do differ in terms of how long they can last without oxygen). But if I want to know how many pounds human beings can bench press, there is wide variation—I'd need to measure a large cross-section of different people to obtain a range and a stable average. I'd want to sample large people, small people, fat people, skinny people, men, women, children, body builders and couch potatoes, people taking anabolic steroids, and teetotalers. There are probably other factors that matter, such as how much sleep the person got the night before testing, how long it's been since they ate, whether they're angry or calm, and so on. Then there are things that we think don't matter at all: whether the air-traffic controller at the St. Hubert Airport in Quebec is male or female that day; whether a random customer at a restaurant in Aberdeen was served in a timely fashion or not. These things may make a difference to other things we're measuring (latent sexism in the air-travel industry; customer satisfaction at Northwestern dining establishments) but not to bench pressing.

The job of the statistician is to formulate an inventory of all those things that matter in order to obtain a representative sample. Researchers have to avoid the tendency to capture variables that are easy to identify or collect data on—sometimes the things that matter are not obvious or are difficult to measure. As Galileo Galilei said, the job of the scientist is to measure what is measurable and to render measurable that which is not. That is, some of the most creative acts in science involve figuring out how to measure something that makes a difference, that no one had figured out how to measure before.

But even measuring and trying to control variables that you know about poses challenges. Suppose you want to study current attitudes about climate change in the United States. You've been given a small sum of money to hire helpers and to buy a statistics program to run on your computer. You happen to live in San Francisco, and so you decide to study there. Already you're in trouble: San Francisco is not representative of the rest of the state of California, let alone the United States. Realizing this, you decide to do your polling in August, because studies show that this is peak tourist season—people from all over the country come to San Francisco then, so (you think) you'll be able to get a cross-section of Americans after all.

But wait: Are the kinds of people who visit San Francisco representative of others? You'd skew toward people who can afford to travel there, and people who want to spend their vacation in a city, as opposed to, say, a national park. (You might also skew toward liberals, because San Francisco is a famously liberal city.)

So you decide that you can't afford to study U.S. attitudes, but you can study attitudes about climate change among San Franciscans. You set up your helpers in Union Square and stop passersby with a short questionnaire. You instruct them to seek out people of different ages, ethnicities, styles of dress, with and without tattoos—in short, a cross-section with variability. But you're still in trouble. You're unlikely to encounter the bedridden, mothers with small children, shift workers who sleep during the day, and the hundreds of thousands of San Francisco residents who, for various reasons, don't come to Union Square, a part of town known for its expensive shops and restaurants. Sending half your helpers down to the Mission District helps solve the problem of different socioeconomic statuses being represented, but it doesn't solve your other problems. The test of a random sample is this: Does everyone or -thing in the whole group have an equal chance of being measured by you and your team? Here the answer is clearly no.

So you conduct a *stratified* random sample. That is, you identify the different strata or subgroups of interest and draw people from them in proportion to the whole population. You do some research about climate change and discover that attitudes do not seem to fall upon racial lines, so you don't need to create subsamples based on race. That's just as well because it can be difficult, or offensive, to make assumptions about race, and what do you do with people of mixed race? Put them in one category or another, or create a new category entirely? But then what? A category for Americans who identify as black-white, black-Hispanic, Asian-Persian, etc.? The categories then may become so specific as to create trouble for your analysis because you have too many distinct groups. Another hurdle: You want age variability, but people don't always feel comfortable telling you their age. You can choose people who are obviously under forty and obviously over forty, but then you miss people who are in their late thirties and early forties.

To even out the problem of people who aren't out and about during the day, you decide to go to people's homes door-to-door. But if you go during the day, you miss working people; if you go during the evening, you miss nightclubbers, shift workers, nighttime church-service attendees, moviegoers, and restaurant-goers. Once you've stratified, how do you get a random sample within your subgroups? All the same problems described above still persist— creating the subgroups doesn't solve the problem that within the subgroup you still have to try to find a fair representation of all the *other* factors that might make a difference to your data. It starts to feel like we will have to sample all the rocks on the moon to get a good analysis.

Don't throw in the trowel—er, towel. Stratified random sampling is better than non-stratified. If you take a poll of college students drawn at random to describe their college experience, you may end up with a sample of students only from large state colleges—a random sample is more likely to choose them because there are so many more of them. If the college experience is radically different

"After sampling every bird that frequents the sidewalk outside this building, we've concluded that what birds really love is bagels!"

at small, private liberal-arts colleges, you need to ensure that your sample includes students from them, and so your stratified sample will ensure you ask people from different-sized schools. Random sampling should be distinguished from convenience or quota sampling—the practice of just talking to people whom you know or people on the street who look like they'll be helpful. Without a random sample, your poll is very likely to be biased.

Collecting data through sampling therefore becomes a never-ending battle to avoid sources of bias. And the researcher is never entirely successful. Every time we read a result in the newspaper that 71 percent of the British are in favor of something, we should reflexively ask, "Yes, but 71 percent of *which* British?"

Add to this the fact that any questions we ask of people are only a sample of all the possible questions we could be asking them, and their answers may only be a sample of what their complex attitudes and experiences are. To make matters worse, they may or may not understand what we're asking, and they may be distracted while they're answering. And, more frequently than pollsters like to admit, people sometimes give an intentionally wrong answer. Humans are a social species; many try to avoid confrontation and want to please, and so give the answer they think the pollster wants to hear. On the other hand, there are disenfranchised members of society and nonconformists who will answer falsely just to shock the pollster, or as a way to try on a rebellious persona to see if it feels good to shock and challenge.

Achieving an unbiased sample isn't easy. When hearing a new statistic, ask, "What biases might have crept in during the sampling?"

Samples give us estimates of something, and they will almost always deviate from the true number by some amount, large or small, and that is the margin of error. Think of this as the price you pay for not hearing from everyone in the population under study, or sampling every moon rock. Of course, there can be errors even if you've interviewed or measured every single person in the population, due to flaws or biases in the measurement device. The margin of error does not address underlying flaws in the research, only the degree of error in the sampling procedure. But ignoring those deeper possible flaws for the moment, there is another measurement or statistic that accompanies any rigorously defined sample: the confidence interval.

The margin of error is how accurate the results are, and the confidence interval is how confident you are that your estimate falls within the margin of error. For example, in a standard two-way poll (in which there are two possibilities for respondents), a random sample of 1,067 American adults will produce a margin of error around 3 percent in either direction (we write ±3%). So if a poll shows that 45 percent of Americans support Candidate A, and 47 percent support Candidate B, the true number is somewhere between 42 and 48 percent for A, and between 44 and 50 percent for B. Note that these ranges overlap. What this means is that the two-percentage-point difference between Candidate A and Candidate B is within the margin of error: We can't say that one of them is truly ahead of the other, and so the election is too close to call.

How confident are we that the margin of error is 3 percent and not more? We calculate a confidence interval. In the particular case I mentioned, I reported the 95 percent confidence interval. That means that if we were to conduct the poll a hundred times, using the same sampling methods, ninety-five out of those hundred times the interval we obtain will contain the true value. Five times out of a hundred, we'd obtain a value outside that range. The confidence interval doesn't tell us how far out of the range—it could be a small difference or a large one; there are other statistics to help with that.

We can set the confidence level to any value we like, but 95 percent is typical. To achieve a narrower confidence interval you can do one of two things: *increase* the sample size for a given confidence level; or, for a given sample size, *decrease* the confidence level. For a given sample size, changing your confidence level from 95 to 99 will increase the size of the interval. In most instances the added expense or inconvenience isn't worth it, given that a variety of external circumstances can change people's minds the following day or week anyway.

Note that for very large populations—like that of the United States—we only need to sample a very small percentage, in this case, less than .0005 percent. But for smaller populations—like that of a corporation or school—we require a much larger percentage. For a company with 10,000 employees, we'd need to sample 964 (almost 10 percent) to obtain the 3 percent margin with 95 percent confidence, and for a company of 1,000 employees, we'd need to sample nearly 600 (60 percent).

Margin of error and confidence interval apply to sampling of any kind, not just samples with people: sampling the proportion of electric cars in a city, of malignant cells in the pancreas, or of mercury in fish at the supermarket. In the figure on page 84, margin of error and sample size are shown for a confidence interval of 95 percent.

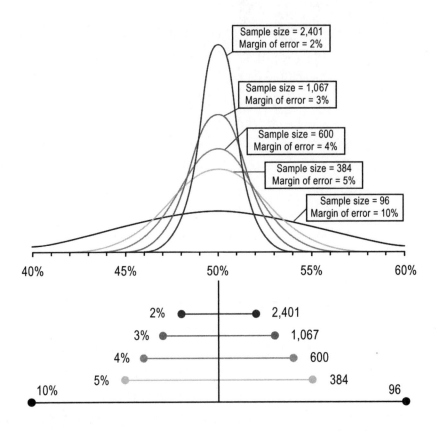

The formula for calculating the margin of error (and confidence interval) is in the notes at the end of the book, and there are many online calculators to help. If you see a statistic quoted and no margin of error is given, you can calculate the margin yourself, knowing the number of people who were surveyed. You'll find that in many cases, the reporter or polling organization doesn't provide this information. This is like a graph without axes—you can lie with statistics very easily by failing to report the margin of error or confidence interval. Here's one: My dog Shadow is the leading gubernatorial candidate in the state of Mississippi, with 76 percent of voters favoring him over other candidates (with an unreported margin of error of ±76 percent; vote for Shadow!!!).

Sampling Biases

While trying to obtain a random sample, researchers sometimes make errors in judgment about whether every person or thing is equally likely to be sampled.

An infamous error was made in the 1936 U.S. presidential election. The *Literary Digest* conducted a poll and concluded that Republican Alf Landon would win over the incumbent Democrat, President Roosevelt. The *Digest* had polled people who were magazine readers, car owners, or telephone customers, not a random sample. The conventional explanation, cited in many scholarly and popular publications, is that in 1936, this skewed heavily toward the wealthy, who were more likely to vote Republican. In fact, according to a poll conducted by George Gallup in 1937, this conventional explanation is wrong—car and telephone owners were more likely to back Roosevelt. The bias occurred in that Roosevelt backers were far less likely to participate in the poll. This sampling bias was recognized by Gallup, who conducted his own poll using a random sample, and correctly predicted the outcome. The Gallup poll was born. And it became the gold standard for political polling until it misidentified the winner in the 2012 U.S. presidential election. An investigation uncovered serious flaws in their sampling procedures, ironically involving telephone owners.

Just as polls based on telephone directories skewed toward the wealthy in the 1930s and 1940s, now landline sampling skews toward older people. All phone sampling assumes that people who own phones are representative of the population at large; they may or may not be. Many Silicon Valley workers use Internet applications for their conversations, and so phone sampling may under-represent high-tech individuals.

If you want to lie with statistics and cover your tracks, take the average height of people near the basketball court; ask about income by sampling near the unemployment office; estimate statewide inci-dence of lung cancer by sampling only near a smelting plant. If you don't disclose how you selected your sample, no one will know.

Participation Bias

Those who are willing to participate in a study and those who are not may differ along important dimensions such as political views, personalities, and incomes. Similarly, those who answer a recruit-ment notice—those who volunteer to be in your study—may show a bias toward or against the thing you're interested in. If you're try-ing to recruit the "average" person in your study, you may bias par-ticipation merely by telling them ahead of time what the study is about. A study about sexual attitudes will skew toward those more willing to disclose those attitudes and against the shy and prudish. A study about political attitudes will skew toward those who are willing to discuss them. For this reason, many questionnaires, sur-veys, and psychological studies don't indicate ahead of time what the research question is, or they disguise the true purpose of the study with a set of irrelevant questions that the researcher isn't interested in.

The people who complete a study may well be different from

those who stop before it's over. Some of the people you contact simply won't respond. This can create a bias when the types of people who respond to your survey are different from the ones who don't, forming a special kind of sampling bias called non-response error.

Let's say you work for Harvard University and you want to show that your graduates tend to earn large salaries just two years after graduation. You send out a questionnaire to everyone in the graduating class. Already you're in trouble: People who have moved without telling Harvard where they went, who are in prison, or who are homeless won't receive your survey. Then, among the ones who respond, those who have high incomes and good feelings about what Harvard did for them might be more likely to fill out the survey than those who are jobless and resentful. The people you don't hear from contribute to non-response error, sometimes in systematic ways that distort the data.

If your goal in conducting the Harvard income-after-two-years survey is to show that a Harvard education yields a high salary, this survey may help you show that to most people. But the critical thinker will realize that the kinds of people who attend Harvard are not the same as the average person. They tend to come from higher-income families, and this is correlated with a student's future earnings. Harvard students tend to be go-getters. They might have earned as high a salary if they had attended a college with a lesser reputation, or even no college at all. (Mark Zuckerberg, Matt Damon, and Bill Gates are financially successful people who dropped out of Harvard.)

If you simply can't reach some segment of the population, such as military personnel stationed overseas, or the homeless and institutionalized, this sampling bias is called *coverage error* because some members of the population from which you want to sample cannot be reached and therefore have no chance of being selected.

If you're trying to figure out what proportion of jelly beans in a jar are red, orange, and blue, you may not be able to get to the bottom of the jar. Biopsies of organs are often limited to where the surgeon can collect material, and this is not necessarily a representative sample of the organ. In psychological studies, experimental subjects are often college undergraduates, who are not representative of the general population. There is a great diversity of people in this country, with differing attitudes, opinions, politics, experiences, and lifestyles. Although it would be a mistake to say that all college students are similar, it would be equally mistaken to say that they represent the rest of the population accurately.

Reporting Bias

People sometimes lie when asked their opinions. A Harvard graduate may overstate her income in order to appear more successful than she is, or may report what she thinks she should have made if it weren't for extenuating circumstances. Of course, she may understate as well so that the Harvard Alumni Association won't hit her up for a big donation. These biases may or may not cancel each other out. The average we end up with in a survey of Harvard graduates' salaries is only the average of what they reported, not what they actually earn. The wealthy may not have a very good idea of their annual income because it is not all salary—it includes a great many other things that vary from year to year, such as income from investments, dividends, bonuses, royalties, etc.

Maybe you ask people if they've cheated on an exam or on their taxes. They may not believe that your survey is truly confidential and so may not want to report their behavior truthfully. (This is a problem with estimating how many illegal immigrants in the U.S. require health care or are crime victims; many are afraid to go to hospitals and police stations for fear of being reported to immigration authorities.)

Suppose you want to know what magazines people read. You could ask them. But they might want to make a good impression on you. Or they might want to think of themselves as more refined in their tastes than they actually are. You may find that a great many more people report reading the *New Yorker* or the *Atlantic* than sales indicate, and a great many fewer people report reading *Us Weekly* and the *National Enquirer*. People don't always tell the truth in surveys. So here, you're not actually measuring what they read, you're measuring snobbery.

So you come up with a plan: You'll go to people's houses and see what magazines they actually have in their living rooms. But this too is biased: It doesn't tell you what they actually read, it only tells you what they choose to keep after they've read it, or choose to display for impression management. Knowing what magazines people read is harder to measure than knowing what magazines people *buy* (or display). But it's an important distinction, especially for advertisers.

What factors underlie whether an individual identifies as multiracial? If they were raised in a single racial community, they may be less inclined to think of themselves as mixed race. If they experienced discrimination, they may be more inclined. *We* might define multiraciality precisely, but it doesn't mean that people will report it the way we want them to.

Lack of Standardization

Measurements must be standardized. There must be clear, replicable, and precise procedures for collecting data so that each person who collects it does it in the same way. Each person who is counting has to count in the same way. Take Gleason grading of tumors—it is only relatively standardized, meaning that you can get different Gleason scores, and hence cancer stage labels, from different pathologists. (In Gleason scoring, a sample of prostate tissue is examined under a microscope and assigned a score from 2 to 10 to indicate how likely it is that a tumor will spread.) Psychiatrists differ in their opinions about whether a certain patient has schizophrenia or not. Statisticians disagree about what constitutes a sufficient demonstration of psychic phenomena. Pathology, psychiatry, parapsychology, and other fields strive to create well-defined procedures that anyone can follow and obtain the same results, but in almost all measurements, there are ambiguities and room for differences of opinion. If you are asked to weigh yourself, do you do so with or without clothes on, with or without your wallet in your pocket? If you're asked to take the temperature of a steak on the grill, do you measure it in one spot or in several and take the average?

Measurement Error

Participants may not understand a question the way the researcher thought they would; they may fill in the wrong bubble on a survey, or in a variety of unanticipated ways, they may not give the answer that they intended. Measurement error occurs in every measurement, in every scientific field. Physicists at CERN reported that they had measured neutrinos traveling faster than the speed of light, a finding that would have been among the most important of the last hundred years. They reported later that they had made an error in measurement.

Measurement error turns up whenever we quantify anything. The 2000 U.S. presidential election came down to measurement error (and to unsuccessfully recording people's intentions): Different teams of officials, counting the same ballots, came up with different numbers. Part of this was due to disagreements over how to count a dimpled chad, a hanging chad, etc.—problems of definition—but even when strict guidelines were put in place, differences in the count still showed up.

We've all experienced this: When counting pennies in our penny jar, we get different totals if we count twice. When standing on a bathroom scale three times in a row, we get different weights. When measuring the size of a room in your house, you may get slightly different lengths each time you measure. These are explainable occurrences: The springs in your scale are imperfect mechanical devices. You hold the tape measure differently each time you use it, it slips from its resting point just slightly, you read the sixteenths of an inch incorrectly, or the tape measure isn't long enough to measure the whole room so you have to mark a spot on the floor and take the measurement in two or three pieces, adding to the possibility of error. The measurement tool itself could have variability (indeed, measurement devices have accuracy specifications attached to them, and the higher-priced the device, the more accurate it tends to be). Your bathroom scale may only be accurate to within half a pound, a postal scale within half an ounce (one thirty-second of a pound).

A 1960 U.S. Census study recorded sixty-two women aged fifteen to nineteen with twelve or more children, and a large number of fourteen-year-old widows. Common sense tells us that there can't be many fifteen- to nineteen-year-olds with twelve children, and fourteen-year-old widows are very uncommon. Someone made an error here. Some census-takers might have filled in the wrong box on a form, accidentally or on purpose to avoid having to conduct time-consuming interviews. Or maybe an impatient (or impish) group of responders to the survey made up outlandish stories and the census-takers didn't notice.

In 2015 the New England Patriots were accused of tampering with their footballs, deflating them to make them easier to catch. They claimed measurement error as part of their defense. Inflation pressures for the footballs of both teams that day, the Pats and the Indianapolis Colts, were taken after halftime. The Pats' balls were tested first, followed by the Colts'. The Colts' balls would have been in a warm locker room or office longer, giving them more time to warm up and thus increase pressure. A federal district court accepted this, and other testimony, and ruled there was insufficient evidence of tampering.

Measurement error also occurs when the instrument you're using to measure—the scale, ruler, questionnaire, or test—doesn't actually measure what you intended it to measure. Using a yardstick to measure the width of a human hair, or using a questionnaire about depression when what you're really studying is motivation (they may be related but are not identical), can create this sort of error. Tallying which candidates people support financially is not the same as knowing how they'll vote; many people give contributions to several candidates in the same race.

Much ink has been spilled over tests or surveys that purport to show one thing but show another. The IQ test is among the most misinterpreted tests around. It is used to assess people's intelligence, as if intelligence were a single quantity, which it is not—it manifests itself in different forms, such as spatial intelligence, artistic intelligence, mathematical intelligence, and so forth. And IQ tests are known to be biased toward middle-class white people. What we usually want to know when we look at IQ test results is how suitable a person is for a particular school program or job. IQ tests can predict performance in these situations, but probably not because the person with a high IQ score is necessarily more intelligent, but because that person has a history of other advantages (economic, social) that show up in an IQ test.

If the statistic you encounter is based on a survey, try to find out what questions were asked and if these seem reasonable and unbiased to you. For any statistic, try to find out how the subject under study was measured, and if the people who collected the data were skilled in such measurements.

Definitions

How something is defined or categorized can make a big difference in the statistic you end up with. This problem arises in the natural sciences, such as in trying to grade cancer cells or describe rainfall, and in the social sciences, such as when asking people about their opinions or experiences.

Did it rain today in the greater St. Louis area? That depends on how you define rain. If only one drop fell to the ground in the 8,846 square miles that comprise "greater St. Louis" (according to the U.S. Office of Management and Budget), do we say it rained? How many drops have to fall over how large an area and over how long a period of time before we categorize the day as one with rainfall?

The U.S. Bureau of Labor Statistics has two different ways of measuring inflation based on two different definitions. The Personal Consumption Expenditures (PCE) and the Consumer Price Index (CPI) can yield different numbers. If you're comparing two years or two regions of the country, of course you need to ensure that you're using the same index each time. If you simply want to make a case about how inflation rose or fell recently, the unscrupulous statistic user would pick whichever of the two made the most impact, rather than choosing the one that is most appropriate, based on an understanding of their differences.

Or what does it mean to be homeless? Is it someone who is sleeping on the sidewalk or in a car? They may have a home and are not able or choose not to go there. What about a woman living on a friend's couch because she lost her apartment? Or a family who has sold their house and is staying in a hotel for a couple of weeks while they wait for their new house to be ready? A man happily and comfortably living as a squatter in an abandoned warehouse? If we compare homelessness across different cities and states, the various jurisdictions may use different definitions. Even if the definition becomes standardized across jurisdictions, a statistic you encounter may not have defined homelessness the way that you would. One of the barriers to solving "the homelessness problem" in our large cities is that we don't have an agreed-upon definition of what it is or who meets the criteria.

Whenever we encounter a news story based on new research, we need to be alert to how the elements of that research have been defined. We need to judge whether they are acceptable and reasonable. This is particularly critical in topics that are highly politicized, such as abortion, marriage, war, climate change, the minimum wage, or housing policy.

And nothing is more politicized than, well, politics. A definition can be wrangled and twisted to anyone's advantage in public-opinion polling by asking a question just-so. Imagine that you've been hired by a political candidate to collect information on his opponent, Alicia Florrick. Unless Florrick has somehow managed to appeal to everyone on every issue, voters are going to have gripes. So here's what you do: Ask the question "Is there anything at all that you disagree with or disapprove of, in anything the candidate has said, even if you support her?" Now almost everyone will have some gripe, so you can report back to your boss that "81 percent of people disapprove of Florrick." What you've done is collected data on one thing (even a single minor disagreement) and swept it into a pile of similar complaints, rebranding them as "disapproval." It almost sounds fair.

Things That Are Unknowable or Unverifiable

GIGO is a famous saying coined by early computer scientists: garbage in, garbage out. At the time, people would blindly put their trust into anything a computer output indicated because the output had the illusion of precision and certainty. If a statistic is composed of a series of poorly defined measures, guesses, misunderstandings, oversimplifications, mismeasurements, or flawed estimates, the resulting conclusion will be flawed.

Much of what we read should raise our suspicions. Ask yourself: Is it possible that someone can know this? A newspaper reports the proportion of suicides committed by gay and lesbian teenagers. Any

such statistic has to be meaningless, given the difficulties in knowing which deaths are suicides and which corpses belong to gay versus straight individuals. Similarly, the number of deaths from
starvation in a remote area, or the number of people killed in a
genocide during a civil war, should be suspect. This was borne out
by the wildly divergent casualty estimates provided by observers
during the Iraq-Afghanistan-U.S. conflict.

A magazine publisher boasts that the magazine has 2 million
readers. How do they know? They don't. They assume some proportion of every magazine sold is shared with others—what they call
the "pass along" rate. They assume that every magazine bought by
a library is read by a certain number of people. The same applies to
books and e-books. Of course, this varies widely by title. Lots of
people *bought* Stephen Hawking's *A Brief History of Time*. Indeed,
it's said to be the most purchased and least finished book of the last
thirty years. Few probably passed it along, because it looks impressive to have it sitting there in the living room. How many readers
does a magazine or book have? How many listeners does a podcast
have? We don't know. We know how many were sold or down-
loaded, that is all (although recent developments with e-books will
probably be changing that long-standing status quo).

The next time that you read that the average New Zealander
flosses 4.36 times a week (a figure I just made up, but it may be as
accurate as any estimate), ask yourself: How could anyone know
such a thing? What data are they relying on? If there were hidden
cameras in bathrooms, that would be one thing, but more likely, it's
people reporting to a survey taker, and only reporting what they
remember—or want to believe is true, because we are always up
against that.

PROBABILITIES

Did you believe me when I said few people *probably* passed along *A Brief History of Time*? I was using the term loosely, as many of us do, but the topic of mathematical probability confronts the very limits of what we can and cannot know about the world, stretching from the behavior of subatomic particles like quarks and bosons to the likelihood that the world will end in our lifetimes, from people playing a state lottery to trying to predict the weather (two endeavors that may have similar rates of success).

Probabilities allow us to quantify future events and are an important aid to rational decision making. Without them, we can become seduced by anecdotes and stories. You may have heard someone say something like "I'm not going to wear my seat belt because I heard about a guy who died in a car crash *because* he was wearing one. He got trapped in the car and couldn't get out. If he hadn't been wearing the seat belt, he would have been okay."

Well, yes, but we can't look at just one or two stories. What are the relative risks? Although there are a few odd cases where the seat belt *cost* someone's life, you're far more likely to die when *not* wearing one. Probability helps us look at this quantitatively.

We use the word *probability* in different ways to mean different things. It's easy to get swept away thinking that a person means one thing when they mean another, and that confusion can cause us to draw the wrong conclusion.

One kind of probability—*classic probability*—is based on the idea of symmetry and equal likelihood: A die has six sides, a coin has two sides, a roulette wheel has thirty-eight slots (in the United States; thirty-seven slots in Europe). If there is no manufacturing defect or tampering that favors one outcome over another, each outcome is equally likely. So the probability of rolling any particular number on a die is one out of six, of getting heads on a coin toss is one out of two, of getting any particular slot on the roulette wheel is one out of thirty-seven or thirty-eight.

Classic probability is restricted to these kinds of well-defined objects. In the classic case, we know the parameters of the system and thus can calculate the probabilities for the events each system will generate. A second kind of probability arises because in daily life we often want to know something about the likelihood of other events occurring, such as the probability that a drug will work on a patient, or that consumers will prefer one beer to another. In this second case, we need to estimate the parameters of the system because we don't know what those parameters are.

To determine this second kind of probability, we make observations or conduct experiments and count the number of times we get the outcome we want. These are called *frequentist* probabilities. We administer a drug to a group of patients and count how many people get better—that's an experiment, and the probability of the drug working is simply the proportion of people for whom it worked (based on the *frequency* of the desired outcome). If we run the experiment on a large number of people, the results will be close to the true probability, just like public-opinion polling.

Both classic and frequentist probabilities deal with recurring, replicable events and the proportion of the time that you can expect to obtain a particular outcome under substantially the same conditions. (Some hard-liner probabilists contend they have to be *identical* conditions, but I think this takes it too far because in the limit, the universe is never *exactly* the same, due to chance variations.) When you conduct a public-opinion poll by interviewing people at random, you're in effect asking them under identical conditions, even if you ask some today and some tomorrow—provided that some big event that might change their minds didn't occur in between. When a court witness testifies about the probability of a suspect's DNA matching the DNA found on a revolver, she is using *frequentist* probability, because she's essentially counting the number of DNA fragments that match versus the number that don't. Drawing a card from a deck, finding a defective widget on an assembly line, asking people if they like their brand of coffee are all examples of classic or frequentist probabilities that are recurring, replicable events (the card is classic, the widget and coffee are frequentist).

A third kind of probability differs from these first two because it's not obtained from an experiment or a replicable event—rather, it expresses an opinion or degree of belief about how likely a particular event is to occur. This is called *subjective probability* (one type of this is Bayesian probability, after the eighteenth-century statistician Thomas Bayes). When a friend tells you that there's a 50 percent chance that she's going to attend Michael and Julie's party this weekend, she's using Bayesian probability, expressing a strength of belief that she'll go. What will the unemployment rate be next year? We can't use the frequentist method because we can't consider next year's unemployment as a set of observations taken under identical or even similar conditions.

Let's think through an example. When a TV weather reporter says that there is a 30 percent chance of rain tomorrow, she didn't conduct experiments on a bunch of identical days with identical conditions (if such a thing even exists) and then count the outcomes. The 30 percent number expresses her degree of belief (on a scale of one to a hundred) that it will rain, and is meant to inform you about whether you want to go to the trouble of grabbing your galoshes and umbrella.

If the weather reporter is well calibrated, it will rain on exactly 30 percent of the days for which she says there is a 30 percent chance of rain. If it rains on 60 percent of those days, she's underestimated by a large amount. The issue of calibration is relevant only with subjective probabilities.

By the way, getting back to your friend who said there is a 50 percent chance she'll attend a party, a mistake that many non–critical thinkers make is in assuming that if there are two possibilities, they must be equally likely. Cognitive psychologists Amos Tversky and Daniel Kahneman described parties and other scenarios to people in an experiment. At a particular party, for example, people might be told that 70 percent of the guests are writers and 30 percent are engineers. If you bump into someone with a tattoo of Shakespeare, you might correctly assume that person to be one of the writers; if you bump into someone wearing a Maxwell's equations T-shirt, you might correctly assume that they are one of the engineers. But what if you bump into someone at random in the party and you've got nothing to go on—no Shakespeare tattoo, no math T-shirt—what's the probability that this person is an engineer? In Tversky and Kahneman's experiments, people tended to say, "Fifty-fifty," apparently confusing the two possible outcomes with two equally likely outcomes.

Subjective probability is the only kind of probability that we have at our disposal in practical situations in which there is no experiment, no symmetry equation. When a judge instructs the jury to return a verdict if the "preponderance of evidence" points toward the defendant's guilt, this is a subjective probability—each juror needs to decide for themselves whether a preponderance has been reached, weighing the evidence according to their own (and possibly idiosyncratic, not objective) internal standards and beliefs.

When a bookmaker lays odds for a horse race, he is using subjective probability—while it might be informed by data on the horses' track records, health, and the jockeys' history, there is no natural symmetry (meaning it's not a classic probability) and there is no experiment being conducted (meaning it's not a frequentist probability). The same is true for baseball and other sporting events. A bookie might say that the Royals have an 80 percent chance of winning their next game, but he's not using probability in a mathematical sense; this is just a way he—and we—use language to give the patina of numerical precision. The bookie can't turn back the hands of time and watch the Royals play the same game again and again, counting how many times they win it. He might well have crunched numbers or used a computer to inform his estimate, but at the end of the day, the number is just a guess, an indication of his degree of confidence in his prediction. A telltale piece of evidence that this is subjective is that different pundits come up with different answers.

Subjective probabilities are all around us and most of us don't even realize it—we encounter them in newspapers, in the boardroom, and in sports bars. The probability that a rogue nation will set off an atomic bomb in the next twelve months, that interest rates will go up next year, that Italy will win the World Cup, or that soldiers will take a particular hill are all subjective, not frequentist:

They are one-time, nonreplicable events. And the reputations of pundits and forecasters depend on their accuracy.

Combining Probabilities

One of the most important rules in probability is the multiplication rule. If two events are independent—that is, if the outcome of one does not influence the outcome of the other—you obtain the probability of *both* of them happening by multiplying the two probabilities together. The probability of getting heads on a coin toss is one half (because there are only two equally likely possibilities: heads and tails). The probability of getting a heart when drawing from a deck of cards is one-quarter (because there are only four equally likely possibilities: hearts, diamonds, clubs, and spades). If you toss a coin and draw a card, the probability of getting both heads and a heart is calculated by multiplying the two individual probabilities together: $\frac{1}{2} \times \frac{1}{4} = \frac{1}{8}$. This is called a joint probability.

You can satisfy yourself that this is true by listing all possible cases and then counting how many times you get the desired outcome:

Head	Heart		Tail	Heart
Head	Diamond		Tail	Diamond
Head	Club		Tail	Club
Head	Spade		Tail	Spade

I'm ignoring the very rare occasions on which you toss the coin and it lands exactly on its side, or it gets carried off by a seagull while it's in midair, or you have a trick deck of cards with all clubs.

We can similarly ask about the joint probability of three events: getting heads on a coin toss, drawing a heart from a deck of cards, and the next person you meet having the same birthday as you (the probability of that is roughly 1 out of 365.24—although births cluster a bit and some birthdates are more common than others, this is a reasonable approximation).

You may have visited websites where you are asked a series of multiple-choice questions, such as "Which of the following five streets have you lived on?" and "Which of the following five credit cards do you have?" These sites are trying to authenticate you, to be sure that you are who they think you are. They're using the multiplication rule. If you answer six of these questions in a row, each with a probability of only one in five (.2) that you'll get it right, the chances of you getting them right by simply guessing are only $.2 \times .2 \times .2 \times .2 \times .2 \times .2$, or .000064—that's about 6 chances in 100,000. Not as strict as what you find in DNA courtroom testimony, but not bad. (If you're wondering why they don't just ask you a bunch of short-answer, fill-in questions, where you have to provide the entire answer yourself, instead of using multiple choice, it's because there are too many variants of correct answers. Do you refer to your credit card as being with Chase, Chase Bank, or JPMorgan Chase? Did you live on North Sycamore Street, N. Sycamore Street, or N. Sycamore St.? You get the idea.)

When the Probability of Events Is Informed by Other Events

The multiplication rule only applies if the events are independent of one another. What events are not independent? The weather, for example. The probability of it freezing tonight *and* freezing tomorrow night are not independent events—weather patterns tend to remain for more than one day, and although freak freezes are known to occur, your best bet about tomorrow's overnight temperatures is to look at today's. You *could* calculate the number of nights in the year in which temperatures drop below freezing—let's say it's thirty-six where you live—and then state that the probability of a freeze tonight is 36 out of 365, or roughly 10 percent, but that doesn't take the dependencies into account. If you say that the probability of it freezing two nights in a row during winter is 10% × 10% = 1% (following the multiplication rule), you'd be underestimating the probability because the two nights' events are not independent; tomorrow's weather forecast is informed by today's.

The probability of an event can also be informed by the particular sample that you're looking at. The probability of it freezing tonight is obviously affected by the area of the world you're talking about. That probability is higher at the forty-fourth parallel than the tenth. The probability of finding someone over six foot six is greater if you're looking at a basketball practice than at a tavern frequented by jockeys. The subgroup of people or things you're looking at is relevant to your probability estimate.

Conditional Probabilities

Often when looking at statistical claims, we're led astray by examining an entire group of random people when we really should be looking at a subgroup. What is the probability that you have pneumonia? Not very high. But if we know more about you and your particular case, the probability may be higher or lower. This is known as a *conditional probability*.

We frame two different questions:

1. What is the probability that a person drawn at random from the population has pneumonia?
2. What is the probability that a person *not* drawn at random, but one who is exhibiting three symptoms (fever, muscle pain, chest congestion) has pneumonia?

The second question involves a conditional probability. It's called that because we're not looking at every possible condition, only those people who match the condition specified. Without running through the numbers, we can guess that the probability of pneumonia is greater in the second case. Of course, we can frame the question so that the probability of having pneumonia is lower than for a person drawn at random:

1. What is the probability that a person *not* drawn at random, but one who has just tested negative for pneumonia three times in a row, and who has an especially robust immune system, and has just minutes ago finished first place in the New York City Marathon, has pneumonia?

Along the same lines, the probability of you developing lung cancer is not independent of your family history. The probability of a waiter bringing ketchup to your table is not independent of what you ordered. You can calculate the probability of any person selected at random developing lung cancer in the next ten years, or the probability of a waiter bringing ketchup to a table calculated over all tables. But we're in the lucky position of knowing that these events are dependent on other behaviors. This allows us to narrow the population we're studying in order to obtain a more accurate estimate. For example, if your father and mother both had lung cancer, you want to calculate the probability of you contracting lung cancer by looking at other people in this select group, people whose parents had lung cancer. If your parents didn't have lung cancer, you want to look at the relevant subgroup of people who lack a family history of it (and you'll likely come up with a different figure). If you want to know the probability that your waiter will bring you ketchup, you might look at only the tables of those patrons who ordered hamburgers or fries, not those who ordered tuna tartare or apple pie.

Ignoring the dependence of events (assuming independence) can have serious consequences in the legal world. One was the case of Sally Clark, a woman from Essex, U.K., who stood trial for murdering her second child. Her first child had died in infancy, and his death had been attributed to SIDS (sudden infant death syndrome, or crib death). The prosecutors argued that the odds of having two children die of SIDS were so low that she must have murdered the second child. The prosecution's witness, a pediatrician, cited a study that said SIDS occurred in 1 out of 8,543 infant deaths. (Dr. Meadow's expertise in pediatrics does not make him an expert statistician or epidemiologist—this sort of confusion is the basis for many faulty judgments and is discussed in Part 3 of this book; an expert in one domain is not automatically an expert in another, seemingly related, domain.)

Digging deeper, we might question the figure of 8,543 deaths. How do they know that? SIDS is a diagnosis of exclusion—that is, there is no test that medical personnel can perform to conclude a death was by

SIDS. Rather, if doctors are not able to find the cause, and they've ruled out everything else, they label it SIDS. Not being able to find something is not proof that it didn't occur, so it is plausible that some of the deaths attributed to SIDS were actually the result of less mysterious causes, such as poisoning, suffocation, heart defect, etc.

For the sake of argument, however, let's assume that SIDS is the cause of 1 out of 8,543 infant deaths as the expert witness testified. He further testified that the odds of two SIDS deaths occurring in the same family were $\frac{1}{8543} \times \frac{1}{8543}$, or 1 in 73 million. ("Coincidence? I think *not!*" the prosecutor might have shouted during his summation.) This calculation—this application of the multiplication rule—assumes the deaths are independent, but they might not be. Whatever caused Mrs. Clark's first child to die suddenly might be present for both children by virtue of them being in the same household: Two environmental factors associated with SIDS are secondhand smoke and putting a baby to sleep on its stomach. Or perhaps the first child suffered from a congenital defect of some sort; this would have a relatively high probability of appearing in the second child's genome (siblings share 50 percent of their DNA). By this way of thinking, there was a 50 percent chance that the second child would die due to a factor such as this, and so now Mrs. Clark looks a lot less like a child murderer. Eventually, her husband found evidence in the hospital archives that the second child's death had a microbiological cause. Mrs. Clark was acquitted, but only after serving three years in prison for a crime she didn't commit.

There's a special notation for conditional probabilities. The probability of a waiter bringing you ketchup, given that you just ordered a hamburger, is written:

P(ketchup | hamburger)

where the vertical bar | is read as *given*. Note that this notation leaves out a lot of the words from the English-language description, so that the mathematical expression is succinct.

The probability of a waiter bringing you ketchup, given that you just ordered a hamburger *and* you asked for the ketchup, is noted:

P(ketchup | hamburger ∧ asked)

where the ∧ is read as *and*.

Visualizing Conditional Probabilities

The relative incidence of pneumonia in the United States in one year is around 2 percent—six million people out of the 324 million in the country are diagnosed each year (of course there are no doubt many undiagnosed cases, as well as individuals who may have more than one case in a year, but let's ignore these details for now). Therefore the probability of any person drawn at random having pneumonia is approximately 2 percent. But we can home in on a better estimate if we know something about that particular person. If you show up at the doctor's office with coughing, congestion, and a fever, you're no longer a person drawn at random—you're someone in a doctor's office showing these symptoms. You can methodically update your belief that something is true (that you have pneumonia) in light of new evidence. We do this by applying Bayes's rule to calculate a conditional probability: What is the probability that I have pneumonia *given* that I show symptom x? This kind of updating can become increasingly refined the more information you have. What is the probability that I have pneumonia *given* that I have these symptoms, and *given* that I have a family history of it, and *given* that I just spent three days with someone who has it? The probabilities climb higher and higher.

You can calculate the probabilities using the formula for Bayes's rule (found in the Appendix), but an easy way to visualize and compute conditional probabilities is with the fourfold table, describing all possible scenarios: You did or didn't order a hamburger, and you did or didn't receive ketchup:

		Ordered Hamburger	
		YES	NO
Received Ketchup	YES		
	NO		

Then, based on experiments and observation, you fill in the various values, that is, the frequencies of each event. Out of sixteen customers you observed at a restaurant, there was one instance of someone ordering a hamburger with which they received ketchup, and two instances with which they didn't. These become entries in the left-hand column of the table:

		Ordered Hamburger	
		YES	NO
Received Ketchup	YES	1	5
	NO	2	8

Similarly, you found that five people who didn't order a hamburger received ketchup, and eight people did not. These are the entries in the right-hand column.

Next, you sum the rows and columns:

Ordered Hamburger

		YES	NO	
Received Ketchup	YES	1	5	6
	NO	2	8	10
		3	13	16

Now, calculating the probabilities is easy. If you want to know the probability that you received ketchup *given* that you ordered a hamburger, you start with the given. That's the left-hand vertical column.

Ordered Hamburger

		YES	NO	
Received Ketchup	YES	1	5	6
	NO	2	8	10
		3	13	16

Three people ordered hamburgers altogether—that's the total at the bottom of the column. Now what is the probability of receiving ketchup *given* you ordered a hamburger? We look now at the "YES received ketchup" square in the "YES ordered hamburger" column, and that number is 1. The conditional probability, P(ketchup|hamburger) is then just one out of three. And you can visualize the logic: three people ordered a hamburger; one of them got ketchup and two didn't. We ignore the right-hand column for this calculation.

We can use this to calculate any conditional probability, including the probability of receiving ketchup if you *didn't* order a hamburger: Thirteen people didn't order a hamburger, five of them got ketchup, so the probability is five out of thirteen, or about 38 percent. In this particular restaurant, you're more likely to get ketchup if you didn't order a hamburger than if you did. (Now fire up your critical thinking. How could this be? Maybe the data are driven by people who ordered fries. Maybe all the hamburgers served already have ketchup on them.)

Medical Decision Making

This way of visualizing conditional probabilities is useful for medical decision making. If you take a medical test, and it says you have some disease, what is the probability you actually have the disease? It's not 100 percent, because the tests are not perfect—they produce false positives (reporting that you have the disease when you don't) and false negatives (reporting that you don't have the disease when you do).

The probability that a woman has breast cancer is 0.8 percent. If she has breast cancer, the probability that a mammogram will indicate it is only 90 percent because the test isn't perfect and it misses some cases. If the woman does not have breast cancer, the probability of a positive result is 7 percent. Now, suppose a woman, drawn at random, has a positive result—what is the probability that she actually has breast cancer?

We start by drawing a fourfold table and filling in the possibilities: The woman actually has breast cancer or doesn't, and the test can report that she does or that she doesn't. To make the numbers work out easily—to make sure we're dealing with whole numbers—let's assume we're talking about 10,000 women.

That's the total population, and so that number goes in the lower right-hand corner of the figure, outside the boxes.

Test Result

		YES	NO	
Actually Has Breast Cancer	**YES**			
	NO			
				10,000

Unlike the hamburger-ketchup example, we fill in the margins first, because that's the information we were given. The probability of breast cancer is 0.8 percent, or 80 out of 10,000 people. That number goes in the margin of the top row. (We don't yet know how to fill in the boxes, but we will in a second.) And because the row has to add up to 10,000, we know that the margin for the bottom row has to equal

10,000 − 80 = 9,920.

Test Result

Actually Has Breast Cancer		YES	NO	
	YES			80
	NO			9,920
				10,000

We were told that the probability that the test will show a positive *if* breast cancer exists is 90 percent. Because probabilities have to add up to 100 percent, the probability that the test will *not* show a positive result if breast cancer exists has to be 100 percent – 90 percent, or 10 percent. For the eighty women who actually have breast cancer (the margin for the top row), we now know that 90 percent of them will have a positive test result (90% of 80 = 72) and 10 percent will have a negative result (10% of 80 = 8). This is all we need to know how to fill in the boxes on the top row.

Test Result

Actually Has Breast Cancer		YES	NO	
	YES	72	8	80
	NO			9,920
				10,000

We're not yet ready to calculate the answer to questions such as "What is the probability that I have breast cancer given that I had a positive test result?" because we need to know how many people will have a positive test result. The missing piece of the puzzle is in the original description: 7 percent of women who don't have breast cancer will still show a positive result. The margin for the lower row tells us 9,920 women don't have breast cancer; 7 percent of them = 694.4. (We'll round to 694.) That means that 9,920 − 694 = 9,226 goes in the lower right square.

		Test Result		
		YES	**NO**	
Actually Has Breast Cancer	**YES**	72	8	80
	NO	694	9,226	9,920
		766	9,234	10,000

Finally, we add up the columns.

If you're among the millions of people who think that having a positive test result means you definitely have the disease, you're wrong. The conditional probability of having breast cancer given a positive test result is the upper left square divided by the left column's margin total, or $72/766$. The good news is that, *even with a positive mammogram, the probability of actually having breast cancer is* 9.4 percent. This is because the disease is relatively rare (less than 1 in 1,000) and the test for it is imperfect.

Test Result

		YES	NO	
Actually Has Breast Cancer	**YES**	72	8	80
	NO	694	9,226	9,920
		766	9,234	10,000

Conditional Probabilities Do Not Work Backward

We're used to certain symmetries in math from grade school: If x = y then y = x. 5 + 7 = 7 + 5. But some concepts don't work that way, as we saw in the discussion above on probability values (if the probability of a false alarm is 10 percent, that doesn't mean that the probability of a hit is 90 percent).

Consider the statistic:

> Ten times as many apples are sold in supermarkets as in roadside stands.

A little reflection should make it apparent that this does not mean you're more likely to find an apple on the day you want one by going to the supermarket: The supermarket may have more than ten times the number of customers as roadside stands have, but even with its greater inventory, it may not keep up with demand. If you see a random person walking down the street with an apple, and you have no information about where they bought it, the probability is higher that they bought it at a supermarket than at a roadside stand.

We can ask, as a conditional probability, what is the probability that this person bought it at a supermarket given that they have an apple?

P(was in a supermarket | found an apple to buy)

It is not the same as what you might want to know if you're craving a Honeycrisp:

P(found an apple to buy | was in a supermarket)

This same asymmetry pops up in various disguises, in all manner of statistics. If you read that more automobile accidents occur at seven p.m. than at seven a.m., what does that mean? Here, the language of the statement itself is ambiguous. It could either mean you're looking at the probability that it was seven p.m. given that an accident occurred, or the probability that an accident occurred given that it was seven p.m. In the first case, you're looking at all accidents and seeing how many were at seven p.m. In the second case, you're looking at how many cars are on the road at seven p.m., and seeing what proportion of them are involved in accidents. What?

Perhaps there are far more cars on the road at seven p.m. than any other time of day, and far fewer accidents per thousand cars. That would yield more accidents at seven p.m. than any other time, simply due to the larger number of vehicles on the road. It is the accident *rate* that helps you determine the safest time to drive.

Similarly, you may have heard that most accidents occur within three miles of home. This isn't because that area is more dangerous per se, it's because the majority of trips people take are short ones, and so the three miles around the home is much more traveled. In most cases, these two different interpretations of the statement will not be equivalent:

P(7 p.m. | accident) ≠ P(accident | 7 p.m.)

The consequences of such confusion are hardly just theoretical: Many court cases have hinged on a misapplication of conditional probabilities, confusing the direction of what is known. A forensics expert may compute, correctly, that the probability of the blood found at the crime scene matching the defendant's blood type by chance is only 1 percent. This is *not* at all the same as saying that there is only a 1 percent chance the defendant is innocent. What? Intuition tricks us again. The forensics expert is telling us the probability of a blood match *given* that the defendant is innocent:

P(blood match | innocence)

Or, in plain language, "the probability that we would find a match if the defendant were actually innocent." That is not the same as the number you really want to know: What is the probability that the defendant is innocent *given* that the blood matched:

P(blood match | innocence) ≠ P(innocence | blood match)

Many innocent citizens have been sent to prison because of this misunderstanding. And many patients have made poor decisions about medical care because they thought, mistakenly, that

P(positive test result | cancer) = P(cancer | positive test result)

And it's not just patients—doctors make this error all the time (in one study 90 percent of doctors treated the two different probabilities the same). The results can be horrible. One surgeon persuaded ninety women to have their healthy breasts removed if they were in a high-risk group. He had noted that 93 percent of breast cancers occurred in women who were in this high-risk group. Given that a woman had breast cancer, there was a 93 percent chance she was in this group: P(high-risk group | breast cancer) = .93. Using a fourfold table for a sample of 1,000 typical women, and adding the additional information that 57 percent of women fall into this high-risk group, and that the probability of a woman having breast cancer is 0.8 percent (as mentioned earlier), we can calculate P(breast cancer | high-risk group), which is the statistic a woman needs to know before consenting to the surgery (numbers are rounded to the nearest integer):

High-Risk Group

		YES	NO	
Actually Has Breast Cancer	**YES**	7	1	8
	NO	563	429	992
		570	430	1,000

The probability that a woman has cancer, given that she is in this high-risk group is not 93 percent, as the surgeon erroneously thought, but only $7/570$, or 1 percent. The surgeon overestimated the cancer risk by nearly one hundred times the actual risk. And the consequences were devastating.

The fourfold tables might feel like a strange little exercise, but actually what you're doing here is scientific and critical thinking, laying out the numbers visually in order to make the computation easier. And the results of those computations allow you to quantify the different parts of the problem, to help you make more rational, evidence-based decisions. They are so powerful, it's surprising that they're not taught to all of us in high school.

Thinking About Statistics and Graphs

Most of us have difficulty figuring probabilities and statistics in our heads and detecting subtle patterns in complex tables of numbers. We prefer vivid pictures, images, and stories. When making decisions, we tend to overweight such images and stories, compared to statistical information. We also tend to misunderstand or misinterpret graphics.

Many of us feel intimidated by numbers and so we blindly accept the numbers we're handed. This can lead to bad decisions and faulty conclusions. We also have a tendency to apply critical thinking only to things we disagree with. In the current information age, pseudofacts masquerade as facts, misinformation can be indistinguishable from true information, and numbers are often at the heart of any important claim or decision. Bad statistics are everywhere. As sociologist Joel Best says, it's not just because the other guys are all lying weasels. Bad statistics are produced by people—often sincere, well-meaning people—who aren't thinking critically about what they're saying.

The same fear of numbers that prevents many people from analyzing statistics prevents them from looking carefully at the numbers in a graph, the axis labels, and the story that they tell. The world is full of coincidences and bizarre things are very likely to happen—but just because two things change together doesn't mean that one caused the other or that they are even related by a hidden *third factor x*. People who are taken in by such associations or coincidences usually have a poor understanding of probability, cause and effect, and the role of randomness in the unfolding of events. Yes, you could spin a story about how the drop in the number of pirates over the last three hundred years and the coinciding rise in global temperatures must surely indicate that pirates were essential to keeping global warming under control. But that's just sloppy thinking, and is a misinterpretation of the evidence. Sometimes the purveyors of this sort of faulty logic know better and hope that you won't notice; sometimes they have been taken in themselves. But now you know better.

PART TWO

EVALUATING WORDS

A lie which is half a truth is ever the blackest of lies.

—ALFRED, LORD TENNYSON

How Do We Know?

We are a storytelling species, and a social species, easily swayed by the opinions of others. We have three ways to acquire information: We can discover it ourselves, we can absorb it implicitly, or we can be told it explicitly. Much of what we know about the world falls in this last category—somewhere along the line, someone told us a fact or we read about it, and so we know it only secondhand. We rely on people with expertise to tell us.

I've never seen an atom of oxygen or a molecule of water, but there is a body of literature describing meticulously conducted experiments that lead me to believe these exist. Similarly, I haven't verified firsthand that Americans landed on the moon, that the speed of light is 186,000 miles per second, that pasteurization really kills bacteria, or that humans normally have twenty-three chromosomes. I don't know firsthand that the elevator in my building has been properly designed and maintained, or that my doctor actually went to medical school. We rely on experts, certifications, licenses, encyclopedias, and textbooks.

But we also need to rely on ourselves, on our own wits and powers of reasoning. Lying weasels who want to separate us from our money, or get us to vote against our own best interests, will try to snow us with pseudo-facts, confuse us with numbers that have no basis, or distract us with information that, upon closer examination, is not actually relevant. They will masquerade as experts.

The antidote to this is to analyze claims we encounter the way we analyze statistics and graphs. The skills necessary should not be beyond the ability of most fourteen-year-olds. They are taught in law schools and schools of journalism, sometimes in business schools and graduate science programs, but rarely to the rest of us, to those who need it most.

If you like watching crime dramas, or reading investigative journalism pieces, many of the skills will be familiar—they resemble the kinds of evaluations that are made during court cases. Judges and juries evaluate competing claims and try to discover the truth within. There are codified rules about what constitutes real evidence; in the United States, documents that haven't been authenticated are generally not allowed, nor is "hearsay" testimony, although there are exceptions.

Suppose someone points you to a website that claims that listening to Mozart music for twenty minutes a day will make us smarter. Another website says it's not true. A big part of the problem here is that the human brain often makes up its mind based on emotional considerations, and then seeks to justify them. And the brain is a very powerful self-justifying machine. It would be nice to believe that all you have to do is listen to beautiful music for twenty minutes to suddenly take your place at the head of the IQ line. It takes effort to evaluate claims like this, probably more time than it would take to listen to *Eine Kleine Nachtmusik*, but it is necessary to avoid drawing incorrect conclusions. Even the smartest of us can be fooled. Steve Jobs delayed treatment for his pancreatic cancer while he followed the advice (given in books and websites) that a change in diet could provide a cure. By the time he realized the diet wasn't working, the cancer had progressed too far to be treated.

Determining the truthfulness or accuracy of a source is not always possible. Consider the epigram opening Part One:

> It ain't what you don't know that gets you into trouble. It's what you know for sure that just ain't so.

I saw this at the opening of the feature film *The Big Short*, which attributed it to Mark Twain, and I felt I had seen it somewhere before; Al Gore also used it in his film *An Inconvenient Truth* nine years earlier with the same attribution. But in fact-checking the *Field Guide,* I could not find any evidence that Twain ever said this. The attribution and quote itself are prime examples of what the quote is trying to warn us against. The directors, writers, and producers of both films didn't do their homework—what they thought they knew for sure turned out not to be true at all.

A little web research pulled up an article in *Forbes* that claims it is a misattribution. The author, Nigel Rees, cites *Respectfully Quoted,* a dictionary of quotations compiled by the U.S. Library of Congress. That book reports various formulations of the remark in *Everybody's Friend, or Josh Billing's Encyclopedia and Proverbial Philosophy of Wit and Humor* (1874). "There you are, you see," writes Rees. "Mark Twain is a better-known humorist than 'Josh Billings' and so the quote drifts towards him."

Rees continues:

And not only him. In a 1984 presidential debate, Walter Mondale came up with this: "I'm reminded a little bit of what Will Rogers once said of Hoover. He said 'It's not what he doesn't know that bothers me, it's what he knows for sure just ain't so.'"

Who's right? With difficult matters such as this, it is often helpful to consult an expert. I asked Gretchen Lieb, a research librarian at Vassar who works as the liaison to the English Department, and who provided this insightful analysis:

> Quotations are tricky things. They're the literary equivalent of statistics, really, in terms of lies, damn lies, etc. Older quotations are almost like translations from another language, too, in terms of being interpretations rather than verbatim, especially in the case of this circle, since these authors wrote in a sort of fantasy dialect, à la Huckleberry Finn, that is difficult to read and downright disturbing to us now in some cases.
>
> I could go check numerous other books of quotations, such as Oxford, etc., but that would be so twentieth century.
>
> Have you come across HathiTrust? It's the corpus of books from research libraries that is behind Google Books, and it's a gold mine, especially for pre-1928 printed materials.
>
> Here's the Josh Billings attribution in "Respectfully Quoted" (we have it as an e-book; I didn't need to walk away from my desk!), and it cites the *Oxford Dictionary of Quotations*, which I tend to use more than Bartlett's:

"The trouble with people is not that they don't know but that they know so much that ain't so." Attributed to Josh Billings (Henry Wheeler Shaw) by *The Oxford Dictionary of Quotations*, 3d ed., p. 491 (1979). Not verified in his writings, although some similar ideas are found in *Everybody's Friend, or Josh Billing's Encyclopedia and Proverbial Philosophy of Wit and Humor* (1874). Original spelling is corrected: "What little I do know I hope I am certain of." (p. 502) "Wisdom don't consist in knowing more that is new, but in knowing less that is false." (p. 430) "I honestly believe it is better to know nothing than to know what ain't so." (p. 286)

By the way, regarding the Walter Mondale attribution to Will Rogers, *Respectfully Quoted* notes that this has not been found in Rogers's work.

Here is a link to Billings's book, where you can search for the phrase "ain't so" and get the idea of what lies therein: http://hdl.handle.net/2027/njp.32101067175438.

Not verifiable. Plus, if you search for Mark Twain, you find that this compendium/encyclopedia writer cites fellow humorist and smartypants Mark Twain as his most trusted correspondent, so they're having a conversation and bouncing clever aphorisms, or as Billings would say, "affurisms," off of each other. Who knows who said what?

I usually roll my eyes when people, especially politicians, quote Mark Twain or Will Rogers, and think to myself, H. L. Mencken, we hardly know you. Critical minds like his are in short supply these days. Poor Josh Billings. Being the second most famous humorist puts you on precarious ground a hundred years later.

So here's an odd case of a quote that appears to have been utterly fabricated, both in its content and its attribution. The basic idea was contained in Billings, although it's not clear if that idea came from him, Twain, or perhaps their buddy Bret Harte. Will Rogers gets put in the mix because, well, it just sort of *sounds* like something he would say.

The quote that opens Part Two was given to me by an acquaintance who misremembered it as:

> The blackest lie is a partial truth that leads you to the wrong conclusion.

It sounded plausible. It would be just like Tennyson to give color to an abstract noun, and to mix the metaphysical with the practical. I only found out the actual quote ("A lie which is half a truth is ever the blackest of lies") when fact-checking for this book. So it goes, as Kurt Vonnegut would say.

In the presence of new or conflicting claims, we can make an informed and evidence-based choice about what is true. We examine the claims for ourselves, and make a decision, acting as our own judge and jury. And as part of the process we usually do well to seek expert opinions. How do we identify them?

IDENTIFYING EXPERTISE

The first thing to do when evaluating a claim by some authority is to ask who or what established their authority. If the authority comes from having been a witness to some event, how credible a witness are they?

Venerable authorities can certainly be wrong. The U.S. government was mistaken about the existence of weapons of mass destruction (WMDs) in Iraq in the early 2000s, and, in a less politically fraught case, scientists thought for many years that humans had twenty-four pairs of chromosomes instead of twenty-three. Looking at what the acknowledged authorities say is not the last step in evaluating claims, but it is a good early step.

Experts talk in two different ways, and it is vital that you know how to tell these apart. In the first way, they review facts and evidence, synthesizing them and forming a conclusion based on the evidence. Along the way, they share with you what the evidence is, why it's relevant, and how it helped them to form their conclusion. This is the way science is supposed to be, the way court trials proceed, and the way the best business decisions, medical diagnoses, and military strategies are made.

The second way experts talk is to just share their opinions. They are human. Like the rest of us, they can be given to stories, to spinning loose threads of their own introspections, what-ifs, and untested ideas. There's nothing wrong with this—some good, testable ideas come from this sort of associative thinking—but it should not be confused with a logical, evidence-based argument. Books and articles for popular audiences by pundits and scientists often contain this kind of rampant speculation, and we buy them because we are impressed by the writer's expertise and rhetorical talent. But properly done, the writer should also lift the veil of authority, let you look behind the curtain, and see at least some of the evidence for yourself.

The term *expert* is normally reserved for people who have undertaken special training, devoted a large amount of time to developing their expertise (e.g., MDs, airline pilots, musicians, or athletes), and whose abilities or knowledge are considered high relative to others'. As such, expertise is a social judgment—we're comparing one person's skill to the skill level of other people in the world. Expertise is relative. Einstein was an expert on physics sixty years ago; he would probably not be considered one if he were still alive today and hadn't added to his knowledge base what Stephen Hawking and so many other physicists now know. Expertise also falls along a continuum. Although John Young is one of only twelve people to have walked on the moon, it would probably not be accurate to say that Captain Young is an *expert* on moonwalking, although he knows more about it than almost anyone else in the world.

Individuals with similar training and levels of expertise will not necessarily agree with one another, and even if they do, these experts are not always right. Many thousands of expert financial analysts make predictions about stock prices that are completely wrong, and some small number of novices turn out to be right. Every British record company famously rejected the Beatles' demo tape, and a young producer with no expertise in popular music, George Martin, signed them to EMI. Xerox PARC, the inventors of the graphical interface computer, didn't see any future for personal computers; Steve Jobs, who had no business experience at all, thought they were wrong. The success of newcomers in these domains is generally understood to be because stock prices and popular taste are highly unpredictable and chaotic. Stuff happens. So it's not that experts are never wrong, it's just that, statistically, they're more likely to be right.

Many inventors and innovators were told "it will never work" by experts, with the Wright brothers and their fellow would-be inventors of motorized flight being an example *par excellence.* The Wright brothers were high school dropouts, with no formal training in aeronautics or physics. Many experts with formal training declared that heavier-than-air flight would never be possible. The Wrights were self-taught, and their perseverance made them de facto experts themselves when they built a functional heavier-than-air airplane, and proved the other experts wrong. Michael Lewis's baseball story *Moneyball* shows how someone can beat the experts by rejecting conventional wisdom and applying logic and statistical analysis to an old problem; Oakland A's manager Billy Beane built a competitive team by using player performance metrics that other teams undervalued, bringing his team to the playoffs two years in a row, and substantially increasing the team's worth.

Experts are often licensed, or hold advanced degrees, or are recognized by other authorities. A Toyota factory-certified mechanic can be considered an expert on Toyotas. The independent or

self-taught mechanic down the street may have just as much expertise, and may well be better and cheaper. It's just that the odds aren't
as good, and it can be difficult to figure that out for yourself. It's just
averages: The average licensed Toyota mechanic is going to know
more about fixing your Toyota than the average independent. Of
course, there are exceptions and you have to bring your own logic to
bear on this. I knew a Mercedes mechanic who worked for a Mercedes dealership for twenty-five years and was among their most
celebrated and top-rated mechanics. He wanted to shorten his com-
mute and be his own boss so he opened up his own shop. His thirty-
five years of experience (by the time I knew him) gave him more
expertise than many of the dealer's younger mechanics. Or another
case: An independent may specialize in certain repairs that the
dealer rarely performs, such as transmission overhaul or reuphol-
stering. You're better off having your differential rebuilt by an independent who does five of those a month than a dealer who probably
only did it once in vocational school. It's like the saying about surgeons: If you need one, you want the doctor who has performed the
same operation you're going to get two hundred times, not once or
twice, no matter how well those couple of operations went.

In science, technology, and medicine, experts' work appears in
peer-reviewed journals (more on those in a moment) or on patents.
They may have been recognized with awards such as a Nobel Prize,
an Order of the British Empire, or a National Medal of Science. In
business, experts may have had experience such as running or starting a company, or amassing a fortune (Warren Buffett, Bill Gates).
Of course, there are smaller distinctions as well—salesperson of the
month, auto mechanic of the year, community "best of" awards
(e.g., best Mexican restaurant, best roofing contractor).

In the arts and humanities, experts may hold university positions or their expertise may be acknowledged by those with university or governmental positions, or by expert panels. These expert panels are typically formed by soliciting advice from previous winners and well-placed scouts—this is how the Nobel and the MacArthur "genius" award nomination and selection panels are constituted.

If people in the arts and humanities have won a prize, such as the Nobel, Pulitzer, Kennedy Center Honors, Polaris Music Prize, Juno, National Book Award, Newbery, or Man Booker Prize, we conclude they are among the experts at their craft. Peer awards are especially useful in judging expertise. ASCAP, an association whose membership is limited to professional songwriters, composers, and music publishers, presents awards voted on by its members; the award is meaningful because those who bestow it constitute a panel of peer experts. The Grammys and the Academy Awards are similarly voted on by peers within the music and film industry, respectively.

You might be thinking, "Wait a minute. There are always elements of politics and personal taste in such awards. My favorite actor/singer/writer/dancer has never won an award, and I'll bet I could find thousands of people who think she's as good as this year's award winner." But that's a different matter. The award system is generally biased toward ensuring that every winner is deserving, which is not the same as saying that every deserving person is a winner. (Recall the discussion of asymmetries earlier.) Those who are recognized by bona fide, respectable awards have usually risen to a level of expertise. (Again, there are exceptions, such as the awarding of a Grammy in 1990, which was later retracted, to lip-syncers Milli Vanilli; or the awarding of a Pulitzer Prize to

Washington Post reporter Janet Cooke, which was withdrawn two days later when it was discovered that the winning story was fraudulent. Novelist Gabriel García Márquez quipped that Cooke should've been awarded the Nobel Prize for *literature*.) When an expert has been found guilty of fraud, does it negate their expertise? Perhaps. It certainly impacts their credibility—now that you know they've lied once, you should be on guard that they may lie again.

Expertise Is Typically Narrow

Dr. Roy Meadow, the pediatrician who testified in the case of the alleged baby killer Sally Clark, had no expertise in medical statistics or epidemiology. He *was* in the medical profession, and the prosecutor who put him on the stand undoubtedly hoped that jurors would assume he had this expertise. William Shockley was awarded a Nobel Prize in physics as one of three inventors of the transistor. Later in life, he promoted strongly racist views that took hold, probably because people assumed that if he was smart enough to win a Nobel, he must know things that others don't. Gordon Shaw, who "discovered" the now widely discredited Mozart effect, was a physicist who lacked training in behavioral science; people probably figured, as they did with Shockley, "He's a physicist—he must be really smart." But intelligence and experience tend to be domain-specific, contrary to the popular belief that intelligence is a single, unified quantity. The best Toyota mechanic in the world may not be able to diagnose what's wrong with your VW, and the best tax attorney may not be able to give the best advice for a breach-of-contract suit. A physicist is probably not the best person to ask about social science.

There's a special place in our hearts (but hopefully not our rational minds) for actors who use their character's image to hawk products. As believable as Sam Waterston was as the trustworthy, ethical district attorney Jack McCoy in *Law & Order*, as an actor he has no special insight into banking and investments, although his commercials for TD Ameritrade were compelling. A generation earlier, Robert Young, who was much loved on TV's *Marcus Welby, M.D.*, did commercials for Sanka. Actors Chris Robinson (*General Hospital*) and Peter Bergman (*All My Children*) hawked Vicks Formula 44; due to FTC regulations (the so-called white coat rule) the actors had to speak a disclaimer that became a widely known catchphrase: "I'm not a doctor, but I play one on TV." Apparently, gullible viewers mistook the actors' authority in a television drama for authority in the real world of medicine.

Source Hierarchy

Some publications are more likely to consult true experts than others, and there exists a hierarchy of information sources. Some sources are simply more consistently reliable than others. In academia, peer-reviewed articles are generally more accurate than books, and books by major publishers are generally more accurate than self-published books (because major publishers are more likely to review and edit the material and have a greater financial incentive to do so). Award-winning newspapers such as the *New York Times*, the *Washington Post*, and the *Wall Street Journal* earned their reputations by being consistently accurate in their coverage of news. They strive to obtain independent verifications for any news story. If one government official tells them something, they get corroboration from another. If a scientist makes a claim, they contact other scientists who don't have any stake in the finding to hear independent opinions. They do make mistakes; even *Times* reporters have been found guilty of fabrications, and the "newspaper of record" prints errata every day. Some people, including Noam Chomsky, have argued that the *Times* is a vessel of propaganda, reporting news about the U.S. government without a proper amount of skepticism. But again, like with auto mechanics, it's a matter of averages—the great majority of what you read in the *New York Times* is likelier to be true than what you read in, for example, the *New York Post*.

Reputable sources want to be certain of facts before publishing them. Many sources have emerged on the Web that do not hold to the same standards, and in some cases, they can break news stories and do so accurately before the more traditional and cautious media do. Many of us learned of Michael Jackson's death from TMZ.com before the traditional media reported it. TMZ was willing to run the story based on less evidence than were the *Los Angeles Times* or NBC. In that particular case, TMZ turned out to be right, but you can't count on this sort of reporting.

A number of celebrity death reports that circulated on Twitter were found to be false. In 2015 alone, these included Carlos Santana, James Earl Jones, Charles Manson, and Jackie Chan. A 2011 fake tweet caused a sell-off of shares for the company Audience, Inc., during which its stock lost 25 percent. Twitter itself saw its shares climb 8 percent—temporarily—after false rumors of a takeover were tweeted, based on a bogus website made to look a great deal like Bloomberg.com's. As the *Wall Street Journal* reported, "The use of false rumors and news reports to manipulate stocks is a centuries-old ruse. The difference today is that the sheer ubiquity and amount of information that courses through markets makes it difficult for traders operating at high speeds to avoid a well-crafted hoax." And it happens to the best of us. Veteran reporter (and part of a team of journalists that was awarded a 1999 Pulitzer Prize) Jonathan Capehart wrote a story for the *Washington Post* based on a tweet by a nonexistent congressman in a nonexistent district.

As with graphs and statistics, we don't want to blindly believe everything we encounter from a good source, nor do we want to automatically reject everything from a questionable source. You shouldn't trust everything you read in the *New York Times*, or reject everything you read on TMZ. Where something appears goes to the credibility of the claim. And, as in a court trial, you don't want to rely on a single witness, you want corroborating evidence.

The Website Domain

The three-digit suffix of the URL indicates the domain. It pays to familiarize yourself with the domains in your country because some of the domains have restrictions, and that can help you establish a site's credibility for a given topic. In the United States, for example, .edu is reserved for nonprofit educational institutions like Stanford.edu (Stanford University); .gov is reserved for official government agencies like CDC.gov (the Centers for Disease Control); .mil for U.S. military organizations, like army.mil. The most famous is probably .com, which is used for commercial enterprises like GeneralMotors.com. Others include .net, .nyc, and .management, which carry no restrictions (!). Caveat emptor. BestElectrical Service.nyc might actually be in New Jersey (and their employees might not even be licensed to work in New York); AlphaAnd OmegaConsulting.management may not know the first or the last thing about management.

Knowing the domain can also help to identify any potential bias. You're more likely to find a neutral report from an educational or nonprofit study (found on a .edu, .gov, or .org site) than on a commercial site, although such sites may also host student blogs and unsupported opinions. And educational and nonprofits are not without bias: They may present information in a way that maximizes donations or public support for their mission. Pfizer.com may be biased in their discussions about drugs made by competing companies, such as GlaxoSmithKline, and Glaxo of course may be biased toward their own products.

Note that you don't always want neutrality. When searching for the owner's manual for your refrigerator, you probably want to visit the (partisan) manufacturer's website (e.g., Frigidaire.com) rather than a site that could be redistributing an outdated or erroneous version of the manual. That .gov site may be biased toward government interests, but a .gov site can give you most accurate info on laws, tax codes, census figures, or how to register your car. CDC .gov and NIH.gov probably have more accurate information about most medical issues than a .com because they have no financial interest.

Who Is Behind It?

Could the website be operating under a name meant to deceive you? The Vitamin E Producers Association might create a website called

NutritionAndYou.info, just to make you think that their claims are unbiased. The president of the grocery chain Whole Foods was caught masquerading as a customer on the Web, touting the quality of his company's groceries. Many rating sites, including Yelp! and Amazon, have found their ratings ballot boxes stuffed by friends and family of the people and products being rated. People are not always who they appear to be on the Web. Just because a website is named U.S. Government Health Service, that doesn't mean it is run by the government; a site named Independent Laboratories doesn't mean that it is independent—it could well be operated by an automobile manufacturer who wants to make its cars look good in not-so-independent tests.

In the 2014 congressional race for Florida's thirteenth district, the local GOP offices created a website with the name of their Democratic opponent, Alex Sink, to trick people into thinking they were giving money to her; in reality, the money went to her opponent, David Jolly. The site, contribute.sinkforcongress2014.com, used Sink's color scheme and featured a smiling photo of her, very similar to the photo on her own site.

Working together, across the aisle to break the gridlock in Washington

Illustration of the website for Democratic Congressional candidate Alex Sink

Illustration of the GOP website used to solicit money for Alex Sink's Republican opponent, David Jolly

The GOP's site does say that the money will be used to defeat Sink, so it's not outright fraud, but let's face it—most people don't take the time to read such things carefully. The most eye-catching parts of the trick site are the large photo of Sink, and the headline Alex Sink | Congress, which strongly implies that the site is *for* Alex Sink, not against her. Not to be outdone, Democrats responded with the same trick, creating the site www.JollyForCongress.com to collect money meant for Sink's rival.

Dentec Safety Specialists and Degil Safety Products are competing companies with similar services and products. Dentec has a website, DentecSafety.com, to market their products, and Degil has a website, DegilSafety.com. However, Degil also registered Dentec Safety.ca to redirect Canadian customers to their own site in order to steal customers. A court case ruled that Degil had to pay Dentec $10,000 and to abandon DentecSafety.ca.

An online vendor operated the website GetCanadaDrugs.com. A court found the site name to be "deceptively misdescriptive." Major points included that the pharmaceutical products did not all originate in Canada, and that only around 5 percent of the website's customers were Canadian. The domain name has now ceased to exist.

Knowing the domain name is helpful but hardly a foolproof verification system. MartinLutherKing.org sounds like a site that would provide information about the great orator and civil rights leader. Because it is a .org site, you might conclude that there is no ulterior motive of profit. The site proclaims that it offers "a true historical examination" of Martin Luther King. Wait a minute. Most people don't begin an utterance by saying, "What I am about to tell you is true." The BBC doesn't begin every news item saying, "This is true." Truth is the default position and we assume others are being truthful with us. An old joke goes, "How do you know that someone is lying to you? Because they begin with the phrase *to be perfectly honest.*" Honest people don't need to preface their remarks this way.

What MartinLutherKing.org contains is a shameful assortment of distortions, anti-Semitic rants, and out-of-context quotes. Who runs the site? Stormfront, a white-supremacy, neo-Nazi hate group. What better way to hide a racist agenda than by promising "the truth" about a great civil rights leader?

Institutional Bias

Are there biases that could affect the way a person or organization structures and presents the information? Does this person or organization have a conflict of interest? A claim about the health value of almonds made by the Almond Growers' Association is not as credible as one made by an independent testing laboratory.

When judging an expert, keep in mind that experts can be biased without even realizing it. For the same tumor, a surgical oncologist may advise surgery, while a radiation oncologist advises radiation and a medical oncologist advises chemotherapy. A psychiatrist may recommend drugs for depression while a psychologist recommends talk therapy. As the old saying goes, if you have a hammer, everything looks like a nail. Who's right? You might have to look at the statistics yourself. Or find a neutral party who has assessed the various possibilities. This is what meta-analyses accomplish in science and medicine. (Or at least they're supposed to.) A meta-analysis is a research technique whereby the results of dozens or hundreds of studies from different labs are analyzed together to determine the weight of evidence supporting a particular claim. It's the reason companies bring in an auditor to look at their accounting records or a financial analyst to decide what a company they seek to buy is really worth. Insiders at the company to be acquired certainly are expert in their own company's financial situation, but they are clearly biased. And not always in the direction you'd think. They may inflate the value of the company if they want to sell, or deflate it if they are worried about a hostile takeover.

Who Links to the Web Page?

A special Google search allows you to see who else links to a web page you land on. Type "link:" followed by the website URL, and Google will return all the sites that link to it. (For example, link:breastcancer .org shows you the two hundred sites that have links to it.) Why might you want to do this? If a consumer protection agency, Better Business Bureau, or other watchdog organization links to a site, you might want to know whether they're praising or condemning it. The page could be the exhibit in a lawsuit. Or it could be linked by an authoritative source, such as the American Cancer Society, as a valuable resource.

Alexa.com tells you about the demographics of site visitors—what country they are from, their educational background, and what sites people visited immediately before visiting the site in question. This information can give you a better picture of who is using the site and a sense of their motivations. A site with drug information that is visited by doctors is probably a more trusted source than one that isn't. Reviews about a local business from people who are from your town are probably more relevant to you than reviews by people who are out of state.

Peer-Reviewed Journals

In peer-reviewed publications, scholars who are at arm's length from one another evaluate a new experiment, report, theory, or claim. They must be expert in the domain they're evaluating. The method is far from foolproof, and peer-reviewed findings are sometimes overturned, or papers retracted. Peer review is not the only system to rely on, but it provides a good foundation in helping us to draw our own conclusions, and like democracy, it's the best such system we have. If something appears in *Nature,* the *Lancet,* or *Cell,* for example, you can be sure it went through rigorous peer review. As when trying to decide whether to trust a tabloid or a serious news organization, the odds are better that a paper published in a peer-reviewed journal is correct.

In a scientific or scholarly article, the report should include footnotes or other citations to peer-reviewed academic literature. Claims should be justified, facts should be documented through citations to respected sources. Ten years ago, it was relatively easy to know whether a journal was reputable, but the lines have become blurred with the proliferation of open-access journals that will print anything for a fee, in a parallel world of pseudo-academia. Reference librarians can help you distinguish the two. Journals that appear on indexes such as PubMed (maintained by the U.S. National Library of Medicine) are selected for their quality; articles you return from a regular search are not. Scholar.Google.com is more restrictive than Google or other search engines, limiting search results to scholarly and academic papers, although it does not vet the journals and many pseudo-academic papers are included. It does do a good job of weeding out things that don't even *resemble* scholarly research, but that's a double-edged sword: That can make it more difficult to know what to believe because so many of the results appear to be valid. Jeffrey Beall, a research librarian at the University of Colorado, Denver, has developed a blacklist of what he calls predatory open-access journals (which often charge high fees to authors). His list has grown from twenty publishers four years ago to more than three hundred today. Other sites exist that help you to vet research papers, such as the Social Science Research Network (ssrn.com).

Regulated Authority

On the Web, there is no central authority to prevent people from making claims that are untrue, no way to shut down an offending site other than going through the costly procedure of obtaining a court injunction.

Off the Web, the lay of the land can be easier to see. Textbooks and encyclopedias undergo careful peer review for accuracy (although that content is sometimes changed under political pressure by school boards and legislatures). Articles at major newspapers in democratic countries are rigorously sourced compared to the untrustworthy government-controlled newspapers of Iran or North Korea, for example. If a drug manufacturer makes a claim, the FDA in the United States (Health Canada in Canada, or similar agencies in other countries) had to certify it. If an ad appears on television, the FTC will investigate claims that it is untrue or misleading (in Canada this is done by the ASC, Advertising Standards Canada; in the U.K. by the ASA, the Advertising Standards Authority; Europe uses a self-regulation organization called the EASA, European Advertising Standards Alliance; many other countries have equivalent mechanisms).

The lying weasels who make fraudulent claims can face punishment, but often the punishment is meager and doesn't serve as much of a deterrent. Energy-drink company Red Bull paid more than $13 million in 2014 to settle a class-action lawsuit for misleading consumers with promises of increased physical and mental performance. In 2015, Target agreed to pay $3.9 million to settle claims that the prices it charged in-store were higher than those it advertised, and that it misrepresented the weights of products. Grocery retailer Whole Foods was similarly charged in 2015 with misrepresenting the weight of its prepackaged food items. Kellogg's paid $4 million to settle a lawsuit over misleading ads that claimed its Frosted Mini-Wheats were "clinically shown to improve kids' attentiveness by 11 percent." While these amounts might sound like a lot to us, to Red Bull ($7.7 billion in revenue for 2014), Kellogg's ($14.6 billion), and Target ($72.6 billion) these fines are little more than a rounding error in their accounting.

Is the Information Current? Discredited?

Unlike books, newspapers, and conventional sources, Web pages seldom carry a date; graphs, charts, and tables don't always reveal the time period they apply to. You can't assume that the "Sales Earnings Year to Date" you read on a Web page today actually covers today in the "To Date," or even that it applies to this year.

Because Web pages are relatively cheap and easy to create, people often abandon them when they're done with them, move on to other projects, or just don't feel like updating them anymore. They become the online equivalent of an abandoned storefront with a lighted neon sign saying "open" when, in fact, the store is closed.

For the various reasons already mentioned—fraud, incompetence, measurement error, interpretation errors—findings and claims become discredited. Individuals who were found guilty in properly conducted trials become exonerated. Vehicle airbags that underwent multiple inspections get recalled. Pundits change their minds. Merely looking at the newness of a site is not enough to ensure that it hasn't been discredited. New sites pop up almost weekly claiming things that have been thoroughly debunked. There are many websites dedicated to exposing urban myths, such as Snopes.com, or to collating retractions, such as RetractionWatch.com.

During the fall of 2015 leading up to the 2016 U.S. presidential elections, a number of people referred to fact-checking websites to verify the claims made by politicians. Politicians have been lying at least since Quintus Cicero advised his brother Marcus to do so in 64 B.C.E. What we have that Cicero didn't is real-time verification. This doesn't mean that all the verifications are accurate or unbiased, dear reader—you still need to make sure that the verifiers don't have a bias for or against a particular candidate or party.

Politifact.com, a site operated by the *Tampa Bay Times,* won a Pulitzer Prize for their reporting, which monitors and fact-checks speeches, public appearances, and interviews by political figures, and uses a six-point meter to rate statements as True, Mostly True, Half True, Mostly False, False, and—at the extreme end of false— Pants on Fire, for statements that are not accurate and completely ridiculous (from the children's playground taunt "Liar, liar, pants on fire"). The *Washington Post* also runs a fact-checking site with ratings from one to four Pinocchios, and awards the prized Geppetto Checkmark for statements and claims that "contain the truth, the whole truth, and nothing but the truth."

As just one example, presidential candidate Donald Trump spoke at a rally on November 21, 2015, in Birmingham, Alabama. To support his position that he would create a Muslim registry in the United States to combat the threat of terrorism from within the country, he recounted watching "thousands and thousands" of Muslims in Jersey City cheering as the World Trade Center came tumbling down on 9/11/2001. ABC News reporter George Stepha- nopoulos confronted Trump the following day on camera, noting that the Jersey City police denied this happened. Trump responded that he saw it on television, with his own eyes, and that it was very well covered. Politifact and the *Washington Post* checked all records of television broadcasts and news reports for the three months following the attacks and found no evidence to support Trump's claim. In fact, Paterson, New Jersey, Muslims had placed a banner on the city's main street that read "The Muslim Community Does Not Support Terrorism." Politifact summarized its findings, writing that Trump's recollection "flies in the face of all evidence we could find. We rate this statement Pants on Fire." The *Washington Post* gave it their Four-Pinocchio rating.

During the same campaign, Hillary Clinton claimed "all of my grandparents" were immigrants. According to Politifact (and based on U.S. census records), only one grandparent was born abroad; three of her four grandparents were born in the United States.

Copied and Pasted, Reposted, Edited?

One way to fool people into thinking that you're really knowledgeable is to find knowledgeable-sounding things on other people's Web pages and post them to your own. While you're at it, why not add your own controversial opinions, which will now be enrobed in the scholarship of someone else, and increase hits to your site? If you've got a certain ideological ax to grind, you can do a hatchet job by editing someone else's carefully supported argument to promote the position opposite of theirs. The burden is on all of us to make sure that we're reading the original, unadulterated information, not someone's mash-up of it.

Supporting Information

Unscrupulous hucksters count on the fact that most people don't bother reading footnotes or tracking down citations. This makes it really easy to lie. Maybe you'd like your website to convince people that your skin cream has been shown to reverse the aging process by ten years. So you write an article and pepper it with footnotes that lead to Web pages that are completely irrelevant to the argument. This will fool a lot of people, because most of them won't actually follow up. Those who do may go no further than seeing that the URL you point to is a relevant site, such as a peer-reviewed journal on aging or on dermatology, even though the article cited says nothing about your product.

Even more diabolically, the citation may actually be peripherally related, but not relevant. You might claim that your skin cream contains Vitamin X and that Vitamin X has been shown to improve skin health and quality. So far, so good. But how? Are the studies of Vitamin X reporting on people who spread it on their skin or people who took it orally? And at what dosage? Does your skin product even have an adequate amount of Vitamin X?

Terminology Pitfalls

You may read on CDC.gov that the incidence of a particular disease is 1 in 10,000 people. But then you stumble on an article at NIH.gov that says the same disease has a prevalence of 1 in 1,000. Is there a misplaced comma here, a typo? Aren't incidence and prevalence the same thing? Actually, they're not. The incidence of a disease is the number of new cases (incidents) that will be reported in a given period of time, for example, in a year. The prevalence is the number of existing cases—the total number of people who have the disease. (And sometimes, people who are afraid of numbers make the at-a-glance error that 1 in 1,000 is less than 1 in 10,000, focusing on that large number with all the zeros instead of the word *in*.)

Take multiple sclerosis (MS), a demyelination disease of the brain and spinal cord. About 10,400 new cases are diagnosed each year in the United States, leading to an incidence of 10,400/322,000,000, or 3.2 cases per 100,000 people—in other words, a 0.0032 percent chance of contracting it. Compare that to the total number of people in the United States who already have it, 400,000, leading to a prevalence rate of 400,000/322,000,000, or 120 cases per 100,000, a 0.12 percent chance of contracting it at some point during your lifetime.

In addition to incidence and prevalence, a third statistic, mortality, is often quoted—the number of people who die from a disease, typically within a particular period of time. For coronary heart disease, 1.1 million new cases are diagnosed each year, 15.5 million Americans currently have it, and 375,295 die from it each year. The probability of being diagnosed with heart disease this year is 0.3 percent, about a hundred times more likely than getting MS; the probability of having it right now is nearly 5 percent, and the probability of dying from it in any given year is 0.1 percent. The probability of dying from it at some point in your life is 20 percent. Of course, as we saw in Part One, all of this applies to the aggregate of all Americans. If we know more about a particular person, such as their family history of heart disease, whether or not they smoke, their weight and age, we can make more refined estimates, using conditional probabilities.

The incidence rate for a disease can be high while the prevalence and mortality rates can be relatively low. The common cold is an example—there are many millions of people who will get a cold during the year (high incidence), but in almost every case it clears up quickly, and so the prevalence—the number of people who have it at any given time—can be low. Some diseases are relatively rare, chronic, and easily managed, so the incidence can be low (not many cases in a year) but the prevalence high (all those cases add up, and people continue to live with the disease) and the mortality is low.

When evaluating evidence, people often ignore the numbers and axis labels, as we've seen, but they also often ignore the verbal descriptors, too. Recall the maps of the United States showing "Crude Birth Rate" in Part One. Did you wonder what "crude birth rate" is? You could imagine that a birth rate might be adjusted by several factors, such as whether the birth is live or not, whether the child survives beyond some period of time, and so on. You might think that because the dictionary definition of the word "crude" is that it is something in a natural or raw state, not yet processed or refined (think crude oil) it must mean the raw, unadulterated, unadjusted number. But it doesn't. Statisticians use the term crude birth rate to count live births (thus it is an adjusted number that subtracts stillborn infants). In trying to decide whether to open a diaper business, you want the crude birth rate, not the total birth rate (because total birth rate includes babies who didn't survive birth).

By the way, a related statistic, the crude death rate, refers to the number of people who die at any age. If you subtract this from the crude birth rate, you get a statistic that public policy makers are (and Thomas Malthus was) very interested in: the RNI, rate of natural increase of a population.

Overlooked, Undervalued
Alternative Explanations

When evaluating a claim or argument, ask yourself if there is another reason—other than the one offered—that could account for the facts or observations that have been reported. There are always alternative explanations; our job is to weigh them against the one(s) offered and determine whether the person drawing the conclusion has drawn the most obvious or likely one.

For example, if you pass a friend in the hall and they don't return your hello, you might conclude that they're mad at you. But alternative explanations are that they didn't see you, were late for a meeting, were preoccupied, were part of a psychology experiment, have taken a vow of silence for an hour, or were temporarily invaded by bodysnatchers. (Or maybe permanently invaded.)

Alternative explanations come up a great deal in pseudoscience and counterknowledge, and they come up often in real science too. Physics researchers at CERN reported that they had discovered neutrinos traveling faster than light. That would have upended a century of Einsteinian theory. It turns out it was just a loose cable in the linear accelerator that caused a measurement error. This underscores the point that a methodological flaw in an extremely complicated experiment is almost always the more likely explanation than something that would cause us to completely rewrite our understanding of the nature of the universe.

Similarly, if a Web page cites experiments showing that a brand-new, previously unheard-of cocktail of vitamins will boost your IQ by twenty points—and the drug companies don't want you to know!—you should wonder how likely it is that nobody else has heard of this, and if an alternative explanation for the claim is simply that someone is trying to make money.

Mentalists, fortune-tellers, and psychics make a lot of money performing seemingly impossible feats of mind reading. One explanation is that they have tapped into a secret, hidden force that goes against everything we know about cause and effect and the nature of space-time. An alternative explanation is that they are magicians, using magic tricks, and simply lying about how they do what they do. Lending credence to the latter view is that professional magicians exist, including James Randi, who, so far, has been able to use clever illusions to duplicate every single feat performed by a mentalist. And often, the magicians—in an effort to discredit the self-proclaimed psychics—will tell you how they did the tricks. In fairness, I suppose that it's possible that it is the *magicians* who are trying to deceive us—they are really psychics who are afraid to reveal their gifts to us (possibly for fear of exploitation, kidnapping, etc.) and they are only *pretending* to use clever illusions. But again, look at the two possibilities: One causes us to throw out everything we know about nature and science, and the other doesn't. Any psychologist, law enforcement officer, businessperson, divorced spouse, foreign service worker, spy, or lawyer can tell you that people lie; they do so for a variety of reasons and with sometimes alarming frequency and alacrity. But if you're facing a claim that seems unlikely, the more likely (alternative) explanation is that the person telling it to you is lying in one way or another.

People who try to predict the future without using psychic powers—military leaders, economists, business strategists—are often wildly off in their predictions because they fail to consider alternative explanations. This has led to a business practice called *scenario planning*—considering all possible outcomes, even those that seem unlikely. This can be very difficult to do, and even experts fail. In 1968, Will and Ariel Durant wrote:

> In the United States the lower birth rate of the Anglo-Saxons has lessened their economic and political power; and the higher birth rate of Roman Catholic families suggests that by the year 2000 the Roman Catholic Church will be the dominant force in national as well as in municipal or state governments.

What they failed to consider was that, during those intervening thirty-two years, many Catholics would leave the Church, and many would use birth control in spite of the Church's prohibitions. Alternative scenarios to their view in 1968 were difficult to imagine.

Social and artistic predictions get upended too: Experts said around the time of the Beatles that "guitar bands are on their way out." The reviews of Beethoven's Fifth Symphony on its debut included a number of negative pronouncements that no one would ever want to hear it again. Science also gets upended. Experts said that fast-moving trains would never work because passengers would die of asphyxiation. Experts thought that light moved through an invisible "ether." Science and life are not static. All we can do is evaluate the weight of evidence and judge for ourselves, using the best tools we have at our disposal. One of those tools that is underused is employing creative thinking to imagine alternatives to the way we've been thinking all along.

Alternative explanations are often critical to legal arguments in criminal trials. The framing effects we saw in Part One, and the failure to understand that conditional probabilities don't work backward, have led to many false convictions.

Proper scientific reasoning entails setting up two (or more) hypotheses and presenting the probabilities for both. In a courtroom, attorneys shouldn't be focusing on the probability of a match, but the probability of two possible scenarios: What is the probability that the blood samples came from the same source, versus the probability that they did not? More to the point, we need to compare the probability of a match given that the subject is guilty with the probability of a match given that the subject is innocent. Or we could compare the probability that the subject is innocent given the data, versus the probability that the subject is guilty given the data. We also need to know the accuracy of the measures. The FBI announced in 2015 that microscopic hair analyses were incorrect 90 percent of the time. Without these pieces of information, it is impossible to decide the case fairly or accurately. That is, if we talk only in terms of a match, we're considering only one-sided evidence, the probability of a match given the hypothesis that the criminal was at the scene of the crime. What we don't know is the probability of a match given alternative hypotheses. And the two need to be compared.

This comes up all the time. In one case in the U.K., the suspect, Dennis Adams, was accused based solely on DNA evidence. The victim failed to pick him out of a lineup, and in court said that Adams did not look like her assailant. The victim added that Adams appeared two decades older than the assailant. In addition, Adams had an alibi for the night in question, which was corroborated by testimony from a third party. The only evidence the prosecution presented at trial was the DNA match. Now, Adams had a brother, whom the DNA would also have matched, but there was no additional evidence that the brother had committed the crime, and so investigators didn't consider the brother. But they also lacked additional evidence against Dennis—the *only* evidence they had was the DNA match. No one in the trial considered the alternative hypothesis that it might have been Dennis's brother. . . . Dennis was convicted both in the original trial and on appeal.

Built by the Ancients to Be Seen from Space

You may have heard the speculation that human life didn't really evolve on Earth, that a race of space aliens came down and seeded the first human life. This by itself is not implausible, it's just that there is no real evidence supporting it. That doesn't mean it's not true, and it doesn't mean we shouldn't look for evidence, but the fact that something *could* be true has limited utility—except perhaps for science fiction.

A 2015 story in the *New York Times* described a mysterious formation on the ground in Kazakhstan that could be seen only from space.

Satellite pictures of a remote and treeless northern steppe reveal colossal earthworks—geometric figures of squares, crosses, lines and rings the size of several football fields, recognizable only from the air and the oldest estimated at 8,000 years old.

The largest, near a Neolithic settlement, is a giant square of 101 raised mounds, its opposite corners connected by a diagonal cross, covering more terrain than the Great Pyramid of Cheops. Another is a kind of three-limbed swastika, its arms ending in zigzags bent counterclockwise.

It's easy to get carried away and imagine that these great designs were a way for ancient humans to signal space aliens, perhaps following strict extraterrestrial instructions. Perhaps it was an ancient spaceship landing pad, or a coded message, something like "Send more food." We humans are built that way—we like to imagine things that are out of the ordinary. We are the storytelling species.

Setting aside the rather obvious fact than any civilization capable of interstellar flight must have had a more efficient communication technology at their disposal than arranging large mounds of dirt on the ground, an alternative explanation exists. Fortunately, the *New York Times* (although not every other outlet that reported the story) provides it, in a quote from Dimitriy Dey, the discoverer of the mysterious stones:

> "I don't think they were meant to be seen from the air," Mr. Dey, 44, said in an interview from his hometown, Kostanay, dismissing outlandish speculations involving aliens and Nazis.

(Long before Hitler, the swastika was an ancient and near-universal design element.) He theorizes that the figures built along straight lines on elevations were "horizontal observatories to track the movements of the rising sun."

An ancient sundial explanation seems more likely than space aliens. It doesn't mean it's true, but part of information literacy and evaluating claims is uncovering plausible alternatives, such as this.

The Missing Control Group

The so-called Mozart effect was discredited because the experiments, showing that listening to Mozart for twenty minutes a day temporarily increased IQ, lacked a control group. That is, one group of people was given Mozart to listen to, and one group of people was given nothing to do. Doing nothing is not an adequate control for doing something, and it turns out if you give people something to do—almost anything—the effect disappears. The Mozart effect wasn't driven by Mozart's music increasing IQ, it was driven by the boredom of doing nothing temporarily decreasing effective IQ.

If you bring twenty people with headaches into a laboratory and give them your new miracle headache drug and ten of them get better, you haven't learned anything. Some headaches are going to get better on their own. How many? We don't know. You'd need to have a control group of people with similar ages and backgrounds, and reporting similar pain. And because just the belief that you might get better can lead to health improvements, you have to give the control group something that enables that belief as much as the medicine under study. Hence the well-known placebo, a pill that is made to look exactly like the miracle headache drug so that no one knows who is receiving what until after the experiment is over.

Malcolm Gladwell spread an invalid conclusion in his book *David and Goliath* by suggesting that people with dyslexia might actually have an advantage in life, leading many parents to believe that their dyslexic children should not receive the educational remedies they need. Gladwell fell for the missing control condition. We don't know how much *more* successful his chosen dyslexics might have been if they had been able to improve their condition.

The missing control group shows up in everyday conversation, where it's harder to spot than in scientific claims, simply because we're not looking for it there. You read—and validate—a new study showing that going to bed every night and waking up every morning at the same time increases productivity and creativity. An artist friend of yours, successful by any measure, counters that she's always just slept whenever she wanted, frequently pulling all-nighters and sometimes sleeping for twenty hours at a time, and she's done just fine. But there's a missing control group. How much *more* productive and creative might she have been with a regular sleep schedule? We don't know.

Two twins were separated at birth and reared apart—one in Nazi Germany and the other in Trinidad and Venezuela. One was raised as a Roman Catholic who joined the Hitler Youth, the other as a Jew. They were reunited twenty-one years later and discovered a bizarre list of similar behaviors that many fascinated people could only attribute to genetics: Both twins scratched their heads with their ring finger, both thought it was funny to sneak up on strangers and sneeze loudly. Both men wore short, neatly trimmed mustaches and rectangular wire-rimmed glasses, rounded at the corner. Both wore blue shirts with epaulets and military-style pockets. Both had the same gait when walking, and the same way of sitting in chairs. Both loved butter and spicy food, flushed the toilet before and after using it, and read the endings of books first. Both wrapped tape around pens and pencils to get a better grip.

Stories like this may cause you to wonder about how our behaviors are influenced by our genes. Or if we're all just automatons, and our actions are predetermined. How else to explain such coincidences?

Well, there are two ways, and they both boil down to a missing control group. A social psychologist might say that the world tends to treat people who look alike in similar ways. The attractive are treated differently from the unattractive, the tall differently from the short. If there's something about your face that just looks honest and free of self-interest, people will treat you differently from how they would if your face suggests otherwise. The brothers' behaviors were shaped by the social world in which they live. We'd need a control group of people who are not related, but who still look astonishingly alike, and were raised separately, in order to draw any firm conclusions about this "natural experiment" of the twins separated at birth.

A statistician or behavioral geneticist would say that of the thousands upon thousands of things that we do, it is likely that any two strangers will share some striking similarities in dress, grooming, penchant for practical jokes, or odd proclivities if you just look long enough and hard enough. Without this control group—bringing strangers together and taking an inventory of their habits—we don't know whether the fascinating story about the twins is driven by genetics or pure chance. It may be that genetics plays a role here, but probably not as large a role as we might think.

Cherry-picking

Our brains are built to make stories as they take in the vastness of the world with billions of events happening every second. There are apt to be some coincidences that don't really mean anything. If a long-lost friend calls just as you're thinking of her, that doesn't mean either of you has psychic powers. If you win at roulette three times in a row, that doesn't mean you're on a streak and should bet your last dollar on the next spin. If your non-certified mechanic fixes your car this time, it doesn't mean he'll be able to do it next time—he may just have gotten lucky.

Say you have a pet hypothesis, for example, that too much Vitamin D causes malaise; you may well find evidence to support that view. But if you're looking only for supporting evidence, you're not doing proper research, because you're ignoring the contradictory evidence—there might be a little of this or a lot, but you don't know because you haven't looked. Colloquially, scientists call this "cherry-picking" the data that suit your hypothesis. Proper research demands that you keep an open mind about any issue, and try to valiantly consider the evidence for and against, and then form an evidence-based (not a "gee, I wish this were so"-based) conclusion.

A companion to the cherry-picking bias is selective windowing. This occurs when the information you have access to is unrepresentative of the whole. If you're looking at a city through the window of a train, you're only seeing a part of that city, and not necessarily a representative part—you have visual access only to the part of the city with train tracks running through it, and whatever biases may attach to that. Trains make noise. Wealthier people usually occupy houses away from the noise, so the people who are left living near the tracks tend to have lower income. If all you know of a city is who lives near the tracks, you are not seeing the entire city.

This is of course related to the discussion in Part One about data gathering (how data are collected), and the importance of obtaining representative samples. We're trying to understand the nature of the world—or at least a new city that the train's passing through—and we want to consider alternative explanations for what we're seeing or being told. A good alternative explanation with broad applicability is that you're only seeing part of the whole picture, and the part you're not seeing may be very different.

Maybe your sister is proudly displaying her five-year-old daughter's painting. It may be magnificent! If you love the painting, frame it! But if you're trying to figure out whether to invest in the child's future as the world's next great painter, you'll want to ask some questions: Who cropped it? Who selected it? How big was the original? How many drawings did the little Picasso make before this one? What came before and what came after? Through selective windowing, you may be seeing part of a series of brilliant drawings or a lovely little piece of a much larger (and unimpressive) work that was identified and cropped by the teacher.

We see selective windowing in headlines too. A headline might announce that "three times more Americans support this new legislation than oppose it." Even if you satisfy yourself, based on the steps in Part One of the *Field Guide*, that the survey was conducted on a representative and sufficiently large sample of Americans, you can't conclude that the majority of Americans support the legislation. It could well be that 1 percent oppose it, 3 percent support it, and 94 percent remain undecided. Translate this same kind of monkey-shines to an election headline stating that five times as many

Republicans support Candidate A than Candidate B for the presidential primaries. That may be true, but the headline might leave out that Candidate C is polling with 80 percent of the vote.

Try tossing a coin ten times. You "know" that it should come up heads half the time. But it probably won't. Even if you toss it 1,000 times, you probably won't get exactly 500 heads. Theoretical probabilities are achieved only with an infinite number of trials. The more coin tosses, the closer you'll get to fifty-fifty heads/tails. It's counterintuitive, but there's a probability very close to 100 percent that somewhere in that sequence you'll get five heads in a row. Why is this so counterintuitive? We didn't evolve brains with a sufficient understanding of what randomness looks like. It's not usually heads-tails-heads-tails, but there are going to be runs (also called streaks) even in a random sequence. This makes it easy to fool someone. Just make a cell phone video recording of yourself tossing a coin 1,000 times in a row. Before each toss, say, "This is going to be the first of five heads in a row." Then, if you get a head, before the next toss, say, "This is going to be the second of five heads in a row." If the next one is a tail, start over. If it's not, before you make the next toss, say, "This is going to be the third of five heads in a row." Then just edit your video so that it only includes those five in a row. No one will be any the wiser! If you want to really impress people, go for ten in a row! (There's roughly a 38 percent chance of that happening in 1,000 tosses. Looking at this another way, if you ask a hundred people in a room to toss a coin five times, there is a 96 percent chance that one of them will get five heads in a row.)

The kinds of experiences that a seventy-five-year-old socialite has with the New York City police department are likely to be very different from those of a sixteen-year-old boy of color; their experiences are selectively windowed by what they see. The sixteen-year-old may report being stopped repeatedly without cause, being racially profiled and treated like a criminal. The seventy-five-year-old may fail to understand how this could be. "All *my* experiences with those officers have been so *nice.*"

Paul McCartney and Dick Clark bought up all the celluloid film of their television appearances in the 1960s, ostensibly so that they could control the way their histories are told. If you're a scholar doing research, or a documentarian looking for archival footage, you're limited to what they choose to release to you. When looking at data or evidence to support a claim, ask yourself if what you're being shown is likely to be representative of the whole picture.

Selective Small Samples

Small samples are usually not representative.

Suppose you're responsible for marketing a new hybrid car. You want to make claims about its fuel efficiency. You send a driver out in the vehicle and find that the car gets eighty miles to the gallon. That looks great—you're done! But maybe you just got lucky. Your competitor does a larger test, sending out five drivers in five vehicles and gets a figure closer to sixty miles per gallon. Who's right? You both are! Suppose that your competitor reported the results like this:

Test 1: 58 mpg
Test 2: 38 mpg
Test 3: 69 mpg
Test 4: 54 mpg
Test 5: 80 mpg

Road conditions, ambient temperature, and driving styles create a great deal of variability. If you were lucky (and your competitor unlucky) your one driver might produce an extreme result that you then report with glee. (And of course, if you want to cherry-pick, you just ignore tests one through four). But if the researcher is pursuing the truth, a larger sample is necessary. An independent lab that tested fifty different excursions might find that the average is something completely different. In general, anomalies are more likely to show up in small samples. *Larger samples more accurately reflect the state of the world.* Statisticians call this *the law of large numbers.*

If you look at births in a small rural hospital over a month and see that 70 percent of the babies born are boys, compared to 51 percent in a large urban hospital, you might think there is something funny going on in the rural hospital. There might be, but that isn't enough evidence to be sure. The small sample is at work again. The large hospital might have reported fifty-one out of a hundred births were boys, and the small might have reported seven out of ten. As with the coin toss mentioned above, the statistical average of fifty-fifty is most recognizable in large samples.

How many is enough? This is a job for a professional statistician, but there are rough-and-ready rules you can use when trying to make sense of what you're reading. For population surveys (e.g., voting preferences, toothpaste preferences, and such), sample-size calculators can readily be found on the Web. For determining the local incidence of something (rates such as how many births are boys, how many times a day the average person reports being hungry) you need to know something about the base rate (or incidence rate) of the thing you're looking for. If a researcher wanted to know how many cases of albinism are occurring in a particular community, and then examined the first 1,000 births and found none, it would be foolish to draw any conclusions: Albinism occurs in only 1 in 17,000 births. One thousand births is too small a sample—"small" relative to the scarcity of the thing you're looking for. On the other hand, if the study was on the incidence of preterm births, 1,000 should be more than enough because they occur in one in nine births.

Statistical Literacy

Consider a street game in which a hat or basket contains three cards, each with two sides: One card is red on both sides, one white on both sides, and one is red on one side and white on the other. The con man draws one card from the hat and shows you one side of it and it is red. He bets you $5 that the other side is also red. He wants you to think that there is a fifty-fifty chance that this is so, so you're willing to bet against him, that is, that the other side is just as likely to be white. You might reason something like this:

He's showing me a red side. So he has pulled either the red-red card or the red-white card. That means that the other side is either red or white with equal probability. I can afford to take this bet because even if I don't win this time, I will win soon after.

Setting aside the gambler's fallacy—many people have lost money by doubling down on roulette only to find out that chance is not a self-correcting process—the con man is relying on you (counting on you?) to make this erroneous assignment of probability, and usually talking fast in order to fractionate your attention. It's helpful to work it out pictorially.

Here are the three cards:

Red	Red	White
White	Red	White

If he is showing you a red side, it could be any one of *three* sides that he's showing you. In two of those cases, the other side is red and in only one case the other side is white. So there is a two in three chance that if he showed you red the other side will be red, not a one in two chance. This is because most of us fail to account for the fact that on the double-red card, he could be showing you *either* side. If you had trouble with this, don't feel bad—similar mistakes were made by mathematical philosopher Gottfried Wilhelm Leibniz and many more recent textbook authors. When evaluating claims based on probabilities, try to understand the underlying model. This can be difficult to do, but if you recognize that probabilities are tricky, and recognize the limitations most of us have in evaluating them, you'll be less likely to be conned. But what if everyone around you is agreeing with something that is, well, wrong? The exquisite new clothes the emperor is wearing, perhaps?

COUNTERKNOWLEDGE

Counterknowledge, a term coined by the U.K. journalist Damian Thompson, is misinformation packaged to look like fact and that some critical mass of people have begun to believe. Examples come from science, current affairs, celebrity gossip, and pseudo-history. It includes claims that lack supporting evidence, and claims for which evidence exists that clearly contradicts them. Take the pseudo-historical claims that the Holocaust, moon landings, or the attacks of September 11, 2001, in the United States never happened, but were part of massive conspiracies. (Counterknowledge doesn't always involve conspiracies—only sometimes.)

Part of what helps counterknowledge spread is the intrigue of imagining *what if it were true?* Again, humans are a storytelling species, and we love a good tale. Counterknowledge initially attracts us with the patina of knowledge by using numbers or statistics, but further examination shows that these have no basis in fact—the purveyors of counterknowledge are hoping you'll be sufficiently impressed (or intimidated) by the presence of numbers that you'll blindly accept them. Or they cite "facts" that are simply untrue.

Damian Thompson tells the story of how these claims can take hold, get under our skin, and cause us to doubt what we know . . .

that is, until we apply a rational analysis. Thompson recalls the time a friend, speaking of the 9/11 attacks in the United States, "grabbed our attention with a plausible-sounding observation: 'Look at the way the towers collapsed vertically, instead of toppling over. Jet fuel wouldn't generate enough heat to melt steel. Only controlled explosions can do that.'"

The anatomy of this counterknowledge goes something like this:

The towers collapsed vertically: This is true. We've seen footage.

If the attack had been carried out the way they told us, you'd expect the building to topple over: This is an unstated, hidden premise. We don't know if this is true. Just because the speaker is asserting it doesn't make it true. This is a claim that requires verification.

Jet fuel wouldn't generate enough heat to melt steel: We don't know if this is true either. And it ignores the fact that other flammables—cleaning products, paint, industrial chemicals—may have existed in the building so that once a fire got going, they added to it.

If you're not a professional structural engineer, you might find these premises plausible. But a little bit of checking reveals that professional structural engineers have found nothing mysterious about the collapse of the towers.

It's important to accept that in complex events, not everything is explainable, because not everything was observed or reported. In the assassination of President John F. Kennedy, the Zapruder film is the only photographic evidence of the sequence of events, and it is incomplete. Shot on a consumer-grade camera, the frame rate is only 18.3 frames per second and it is low-resolution. There are many unanswered questions about the assassination, and indications that evidence was mishandled, many eyewitnesses were never questioned, and many unexplained deaths of people who claimed or were presumed to know what really happened. There may well have been a conspiracy, but the mere fact that there are unanswered questions and inconsistencies is not proof of one. An unexplained headache with blurred vision is not evidence of a rare brain tumor—it is more likely something less dramatic.

Scientists and other rational thinkers distinguish between things that we know are almost certainly true—such as photosynthesis or that the Earth revolves around the sun—and things that are *probably* true, such as that the 9/11 attacks were the result of hijacked airplanes, not a U.S. government plot. There are different amounts of evidence, and different kinds of evidence, weighing in on each of these topics. And a few holes in an account or a theory does not discredit it. A *handful* of unexplained anomalies does not discredit or undermine a well-established theory that is based on *thousands* of pieces of evidence. Yet these anomalies are typically at the heart of all conspiratorial thinking, Holocaust revisionism, anti-evolutionism, and 9/11 conspiracy theories. The difference between a false theory and a true theory is one of probability. Thompson dubs something counterknowledge when it runs contrary to real knowledge and has some social currency.

When Reporters Lead Us Astray

News reporters gather information about important events in two different ways. These two ways are often incompatible with each other, resulting in stories that can mislead the public if the journalists aren't careful.

In *scientific investigation* mode, reporters are in a partnership with scientists—they report on scientific developments and help to translate them into a language that the public can understand, something that most scientists are not good at. The reporter reads about a study in a peer-reviewed journal or press release. By the time a study reaches peer review, usually three to five unbiased and established scientists have reviewed the study and accepted its accuracy and its conclusions. It is not usually the reporter's job to establish the weight of scientific evidence supporting every hypothesis, auxiliary hypothesis, and conclusion; that has already been done by the scientists writing the paper.

Now the job splits off into two kinds of reporters. The serious investigative reporter, such as for the *Washington Post*, or the *Wall Street Journal*, will typically contact a handful of scientists *not associated* with the research to get their opinions. She will seek out opinions that go against the published report. But the vast majority of reporters consider that their work is done if they simply report on the story as it was published, translating it into simpler language.

In *breaking news* mode, reporters try to figure out something that's going on in the world by gathering information from sources—witnesses to events. This can be someone who witnessed a holdup in Detroit or a bombing in Gaza or a buildup of troops in

Crimea. The reporter may have a single eyewitness, or try to corroborate with a second or third. Part of the reporter's job in these cases is to ascertain the veracity and trustworthiness of the witness. Questions such as "Did you see this yourself?" or "Where were you when this happened?" help to do so. You'd be surprised at how often the answer is no, or how often people lie, and it is only through the careful verifications of reporters that inconsistencies come to light.

So in Mode One, journalists report on scientific findings, which themselves are probably based on thousands of observations and a great amount of data. In Mode Two, journalists report on events, which are often based on the accounts of only a few eyewitnesses.

Because reporters have to work in both these modes, they sometimes confuse one for the other. They sometimes forget that the plural of anecdote is not data; that is, a bunch of stories or casual observations do not make science. Tangled in this is our expectation that newspapers should entertain us as we learn, tell us stories. And most good stories show us a chain of actions that can be related in terms of cause and effect. Risky mortgages were repackaged into AAA-rated investment products, and that led to the housing collapse of 2007. Regulators ignored the buildup of debris above the Chinese city of Shenzhen, and in 2015 it collapsed and created an avalanche that toppled thirty-three buildings. These are not scientific experiments, they are events that we try to make sense of, to make stories out of. The burden of proof for news articles and scientific articles is different, but without an explanation, even a tentative one, we don't have much of a story. And newspapers, magazines, books—people—need stories.

This is the core reason why rumors, counterknowledge, and pseudo-facts can be so easily propagated by the media, as when Geraldo Rivera contributed to a national panic about Satanists taking over America in 1987. There have been similar media scares about alien abduction and repressed memories. As Damian Thompson notes, "For a hard-pressed news editor, anguished testimony trumps dry and possibly inconclusive statistics every time."

Perception of Risk

We assume that newspaper space given to crime reporting is a measure of crime rate. Or that the amount of newspaper coverage given over to different causes of death correlates to risk. But assumptions like this are unwise. About five times more people die each year of stomach cancer than of unintentional drowning. But to take just one newspaper, the *Sacramento Bee* reported no stories about stomach cancer in 2014, but three on unintentional drownings. Based on news coverage, you'd think that drowning deaths were far more common than stomach-cancer deaths. Cognitive psychologist Paul Slovic showed that people dramatically overweight the relative risks of things that receive media attention. And part of the calculus for whether something receives media attention is whether or not it makes a good story. A death by drowning is more dramatic, more sudden, and perhaps more preventable than death by stomach cancer—all elements that make for a good, though tragic, tale. So drowning deaths are reported more, leading us to believe, erroneously, that they're more common. Misunderstandings of risk can lead us to ignore or discount evidence we could use to protect ourselves.

"You're awfully cavalier considering how much tsunamis have been in the headlines lately."

Using this principle of misunderstood risk, unscrupulous or simply uninformed amateur statisticians with a media platform can easily bamboozle us into believing many things that are not so.

A front-page headline in the *Times* (U.K.) in 2015 announced that 50 percent of Britons would contract cancer in their lifetimes, up from 33 percent. This could rise to two-thirds of today's children, posing a risk that the National Health Service will be overwhelmed by the number of cancer patients. What does that make you think? That there is a cancer epidemic on the rise? Perhaps something about our modern lifestyle with healthless junk food, radiation-emitting cell phones, carcinogenic cleaning products, and radiation coming through a hole in the ozone layer is suspect. Indeed, this headline could be used to promote an agenda by any number of profit-seeking stakeholders—health food companies, sunblock manufacturers, holistic medicine practitioners, and yoga instructors.

Before you panic, recognize that this figure represents all kinds of cancer, including slow-moving ones like prostate cancer, melanomas that are easily removed, etc. It doesn't mean that everyone who contracts cancer will die. Cancer Research UK (CRUK) reports that the percentage of people beating cancer has doubled since the 1970s, thanks to early detection and improved treatment.

What the headline ignores is that, thanks to advances in medicine, people are living longer. Heart disease is better controlled than ever and deaths from respiratory diseases have decreased dramatically in the last twenty-five years. The main reason why so many people are dying of cancer is that they're not dying of other things first. You have to die of *something*. This idea was contained in the same story in the *Times,* if you read that far (which many of us don't; we just stop at the headline and then fret and worry). Part of what the headline statistic reflects is that cancer is an old-person's disease, and many of us now will live long enough to get it. It is not necessarily a cause for panic. This would be analogous to saying, "Half of all cars in Argentina will suffer complete engine failure during the life the car." Yes, of course—the car has to be put out of service for some reason. It could be a broken axle, a bad collision, a faulty transmission, or an engine failure, but it has to be something.

Persuasion by Association

If you want to snow people with counterknowledge, one effective technique is to get a whole bunch of verifiable facts right and then add only one or two that are untrue. The ones you get right will have the ring of truth to them, and those intrepid Web explorers who seek to verify them will be successful. So you just add one or two untruths to make your point and many people will haplessly go along with you. You persuade by associating bogus facts or counterknowledge with actual facts and actual knowledge.

Consider the following argument:

1. Water is made up of hydrogen and oxygen.
2. The molecular symbol for water is H_2O.
3. Our bodies are made up of more than 60 percent water.
4. Human blood is 92 percent water.
5. The brain is 75 percent water.
6. Many locations in the world have contaminated water.
7. Less than 1 percent of the world's accessible water is drinkable.
8. You can only be sure that the quality of your drinking water is high if you buy bottled water.
9. Leading health researchers recommend drinking bottled water, and the majority drink bottled water themselves.

Assertions one through seven are all true. Assertion eight doesn't follow logically, and assertion nine, well . . . who are the leading health researchers? And what does it mean that they drink bottled water themselves? It could be that at a party, restaurant, or on an airplane, when it is served and there are no alternatives, they'll drink it. Or does it mean that they scrupulously avoid all other forms of water? There is a wide chasm between these two possibilities.

The fact is that bottled water is at best no safer or healthier than most tap water in developed countries, and in some cases less safe because of laxer regulations. This is based on reports by a variety of reputable sources, including the Natural Resources Defense Council, the Mayo Clinic, *Consumer Reports*, and a number of reports in peer-reviewed journals.

Of course, there are exceptions. In New York City; Montreal; Flint, Michigan; and many other older cities, the municipal water supply is carried by lead pipes and the lead can leech into the tap water and cause lead poisoning. Periodic treatment-plant problems lead city governments to impose a temporary advisory on tap water. And when traveling in Third World countries, where regulation and sanitation standards are lower, bottled water may be the best bet. But tap-water standards in industrialized nations are among the most stringent standards in any industry—save your money and skip the plastic bottle. The argument of pseudo-scientific health advocates as typified by the above does not, er, hold water.

PART THREE

EVALUATING THE WORLD

Nature permits us to calculate only probabilities. Yet science has not collapsed.

—RICHARD P. FEYNMAN

How Science Works

The development of critical thinking over many centuries led to a paradigm shift in human thought and history: the scientific revolution. Without its development and practice in cities like Florence, Bologna, Göttingen, Paris, London, and Edinburgh, to name just a handful of great centers of learning, science may not have come to shape our culture, industry, and greatest ambitions as it has. Science is not infallible, of course, but scientific thinking underlies a great deal of what we do and of how we try to decide what is and isn't so. This makes it worth taking a close look behind the curtain to better see how it does what it does. That includes seeing how our imperfect human brains, those of even the most rigorous thinkers, can fool themselves.

Unfortunately, we must also recognize that some researchers make up data. In the most extreme cases, they report data that were never collected from experiments that were never conducted. They get away with it because fraud is relatively rare among researchers and so peer reviewers are not on their guard. In other cases, an investigator changes a few data points to make the data more closely reflect his or her pet hypotheses. In less extreme cases, the investigator omits certain data points because they don't conform to the hypothesis, or selects only cases that he or she knows will contribute favorably to the hypothesis. A case of fraud occurred in 2015 when Dong-Pyou Han, a former biomedical scientist at Iowa State University in Ames, was found to have fabricated and falsified data about a potential HIV vaccine. In an unusual outcome, he didn't just lose his job at the university but was sentenced to almost five years in prison.

The entire controversy about whether the measles, mumps, and rubella (MMR) vaccine causes autism was propagated by Andrew Wakefield in an article with falsified data that has now been retracted—and yet millions of people continue to believe in the connection. In some cases, a researcher will manipulate the data or delete data according to established principles, but fail to report these moves, which makes interpretation and replication more difficult (and which borders on scientific misconduct).

The search for proof, for certainty, drives science, but it also drives our sense of justice and all our judicial systems. Scientific practice has shown us the right way to proceed with this search.

There are two pervasive myths about how science is done. The first is that science is neat and tidy, that scientists never disagree about anything. The second is that a single experiment tells us all we need to know about a phenomenon, that science moves forward in leaps and bounds after every experiment is published. Real science is replete with controversy, doubts, and debates about what we really know. Real scientific knowledge is gradually established through many replications and converging findings. Scientific knowledge comes from amassing large amounts of data from a large number of experiments, performed by multiple laboratories. Any one experiment is just a brick in a large wall. Only when a critical mass of experiments has been completed are we in a position to regard the entire wall of data and draw any firm conclusions.

The unit of currency is not the single experiment, but the meta-analysis. Before scientists reach a consensus about something, there has usually been a meta-analysis, tying together the different pieces of evidence for or against a hypothesis.

If the idea of a meta-analysis versus a single experiment reminds you of the selective windowing and small sample problems mentioned in Part Two, it should. A single experiment, even with a lot of participants or observations, could still just be an anomaly—that eighty miles per gallon you were lucky to get the one time you tested your car. A dozen experiments, conducted at different times and places, give you a better idea of how robust the phenomenon is. The next time you read that a new face cream will make you look twenty years younger, or about a new herbal remedy for the common cold, among the other questions you should ask is whether a meta-analysis supports the claim or whether it's a single study.

Deduction and Induction

Scientific progress depends on two kinds of reasoning. In deduction, we reason from the general to the specific, and if we follow the rules of logic, we can be certain of our conclusion. In induction, we take a set of observations or facts, and try to come up with a general principle that can account for them. This is reasoning from the specific to the general. The conclusion of inductive reasoning is not certain—it is based on our observations and our understanding of the world, and it involves a leap beyond what the data actually tell us.

Probability, as introduced in Part One, is deductive. We work from general information (such as "this is a fair coin") to a specific prediction (the probability of getting three heads in a row). Statistics is inductive. We work from a particular set of observations (such as flipping three heads in a row) to a general statement (about whether the coin is fair or not). Or as another example, we would use probability (deduction) to indicate the likelihood that a particular headache medicine will help you. If your headache didn't go away, we could use statistics (induction) to estimate the likelihood that your pill came from a bad batch.

Induction and deduction don't just apply to numerical things like probability and statistics. Here is an example of deductive logic in words. If the premise (the first statement) is true, the conclusion must be also:

> Gabriel García Márquez is a human.
> All humans are mortal.
> Therefore (this is the deductive conclusion) Gabriel García
> Márquez is mortal.

1. Some automobiles are Fords.

2. All Fords are automobiles.

3. The guy who played Han Solo is an automobile!

The type of deductive argument about Márquez is called a syllogism. In syllogisms, it is the *form* of the argument that guarantees that the conclusion follows. You can construct a syllogism with a premise that you know (or think to be) false, but that doesn't invalidate the syllogism—in other words, the logic of the whole thing still holds.

> The moon is made of green cheese.
> Green cheese costs $22.99 per pound.
> Therefore, the moon costs $22.99 per pound.

Now, clearly the moon is *not* made of green cheese, but IF it were, the deduction is logically valid. If it makes you feel better, you can rewrite the syllogism so that this is made explicit:

> IF the moon is made of green cheese
> AND IF green cheese costs $22.99 per pound
> THEN the moon costs $22.99 per pound.

There are several distinct types of deductive arguments, and they're typically taught in philosophy or math classes on formal logic. Another common form involves conditionals. This one is called *modus ponens*. It's easy to remember what it's called with this example (using Poe as in *ponens*):

> If Edgar Allan Poe went to the party, he wore a black cape.
> Edgar Allan Poe went to the party.
> Therefore, he wore a black cape.

Formal logic can take some time to master, because, as with many forms of reasoning, our intuitions fail us. In logic, as in running a race, order matters. Does the following sound like a valid or invalid conclusion?

> If Edgar Allan Poe went to the party, he wore a black cape.
> Edgar Allan Poe wore a black cape.
> Therefore, he went to the party.

While it *might* be true that Poe went to the party, it is not *necessarily* true. He could have worn the cape for another reason (perhaps it was cold, perhaps it was Halloween, perhaps he was acting in a play that required a cape and wanted to get in character). Drawing the conclusion above represents an error of reasoning called the *fallacy of affirming the consequent,* or the *converse error.*

If you have a difficult time remembering what it's called, consider this example:

> If Chuck Taylor is wearing Converse shoes, then his feet are covered.
> Chuck Taylor's feet are covered.
> Therefore, he is wearing Converse shoes.

This reasoning obviously doesn't hold, because wearing Converse shoes is not the only way to have your feet covered—you could be wearing any number of different shoe brands, or have garbage bags on your feet, tied around the ankles.

However, you *can* say with certainty that if Chuck Taylor's feet are not covered, he is not wearing Converse shoes. This is called the *contrapositive* of the first statement.

Logical statements don't work like the minus signs in equations—you can't just negate one side and have it automatically negate the other. You have to memorize these rules. It's somewhat easier to do using quasi-mathematical notation. The statements above can be represented this way, where A stands for any premise, such as "If Chuck Taylor is wearing Converse shoes," or "If the moon is made of green cheese" or "If the Mets win the pennant this year." B is the consequence, such as "then Chuck's feet are covered," or "then the moon should appear green in the night sky" or "I will eat my hat."

Using this generalized notation, we say *If A* as a shorthand for "If A is true." We say *B* or *Not B* as a shorthand for "B is true" or "B is not true." So . . .

If A, then B

A

Therefore, B

In logic books, you may see the word *then* replaced with an arrow (→) and you may see the word *not* replaced with this symbol: ~. You may see the word *therefore* replaced with ∴ as in:

If A → B

A

∴ B

Don't let that disturb you. It's just some people trying to be fancy.

Now there are four possibilities for statements like this: A can be true or not true, and B can be true or not true. Each of the possibilities has a special name.

I. Modus ponens. This is also called affirming the antecedent. "Ante" means before, like when you "ante up" in poker, putting money in the pot before any cards are played.

If A → B

A ∴ B

VALID

Example: If that woman is my sister, then she is younger than I am.

That woman is my sister.

Therefore, she is younger than I am.

2. The contrapositive.

If A → B

~ B ∴ ~ A

Example: If that woman is my sister, then she is younger
than I am.

That woman is not younger than I am.

Therefore, she is not my sister.

3. The converse.

If A → B

B ∴ A

This is a *not* a valid deduction.

Example: If that woman is my sister, then she is younger
than I am.

That woman is younger than I am.

Therefore, she is my sister.

This is invalid because there are many women younger than I am
who are not my sister.

4. The inverse.

If A → B

~A ∴ ~B

This is a *not* a valid deduction.

Example: If that woman is my sister, then she is younger
than I am.
That woman is not my sister.
Therefore, she is not younger than I am.

This is invalid because many women who are not my sister are still younger than I am.

Inductive reasoning is based on there being evidence that suggests the conclusion is true, but does not guarantee it. Unlike deduction, it leads to uncertain but (if properly done) probable conclusions.

An example of induction is:

All mammals we have seen so far have kidneys.
Therefore (this is the inductive step), if we discover a new
mammal, it will probably have kidneys.

Science progresses by a combination of deduction and induction. Without induction, we'd have no hypotheses about the world. We use it all the time in daily life.

Every time I've hired Patrick to do a repair around the
house, he's botched the job.
Therefore, if I hire Patrick to do this next repair, he'll botch
this one too.

Every airline pilot I've met is organized, conscientious, and
meticulous.

Lee is an airline pilot. He has these qualities, and he's also
good at math.

Therefore, all airline pilots are good at math.

Of course, this second example doesn't necessarily follow. We're
making an inference. With what we know about the world, and the
job requirements for being a pilot—plotting courses, estimating the
influence of wind velocity on arrival time, etc.—this seems reasonable. But consider:

Every airline pilot I've met is organized, conscientious, and
meticulous.

Lee is an airline pilot. He has these qualities, and he also
likes photography.

Therefore, all airline pilots like photography.

Here our inference is less certain. Our real-world knowledge
suggests that photography is a personal preference, and it doesn't
necessarily follow that a pilot would enjoy it more or less than a
non-pilot.

The great fictional detective Sherlock Holmes draws conclusions
through clever reasoning, and although he claims to be using
deduction, in fact he's using a different form of reasoning called
abduction. Nearly all of Holmes's conclusions are clever guesses,
based on facts, but not in a way that the conclusion is airtight or
inevitable. In abductive reasoning, we start with a set of observations and then generate a theory that accounts for them. Of the

infinity of different theories that could account for something, we seek the most likely.

For example, Holmes concludes that a supposed suicide was really a murder:

> HOLMES: The wound was on the right side of his head. Van Coon was left-handed. Requires quite a bit of contortion.
> DETECTIVE INSPECTOR DIMMOCK: Left-handed?
> HOLMES: Oh, I'm amazed you didn't notice. All you have to do is look around this flat. Coffee table on the left-hand side; coffee mug handle pointing to the left. Power sockets: habitually used the ones on the left . . . Pen and paper on the left-hand side of the phone because he picked it up with his right and took down messages with his left . . . There's a knife on the breadboard with butter on the right side of the blade because he used it with his left. It's highly unlikely that a left-handed man would shoot himself in the *right* side of his head. Conclusion: Someone broke in here and murdered him . . .
> DIMMOCK: But the gun . . . why—
> HOLMES: He was waiting for the killer. He'd been threatened.

Note that Sherlock uses the phrase *highly unlikely*. This signals that he's not using deduction. And it's not induction because he's not going from the specifics to the general—in a way, he's going from one set of specifics (the observations he makes in the victim's flat) to another specific (ruling it murder rather than suicide). Abduction, my dear Watson.

Arguments

When evidence is offered to support a statement, these combined statements take on a special status—what logicians call an argument. Here, the word *argument* doesn't mean a dispute or disagreement with someone; it means a formal logical system of statements. Arguments have two parts: evidence and a conclusion. The evidence can be one or more statements, or premises. (A statement without evidence, or without a conclusion, is not an argument in this sense of the word.)

Arguments set up a system. We often begin with the conclusion—I know this sounds backward, but it's how we typically speak; we state the conclusion and *then* bring out the evidence.

> Conclusion: Jacques cheats at pool.
> Evidence (or premise): When your back was turned, I saw him move the ball before taking a shot.

Deductive reasoning follows the process in the opposite direction.

> Premise: When your back was turned, I saw him move the ball before taking a shot.
> Conclusion: Jacques cheats at pool.

This is closely related to how scientists talk about the results of experiments, which are a kind of argument, again in two parts.

> Hypothesis = H
> Implication = I

H: There are no black swans.

I: If H is true, then neither I nor anyone else will ever see a
black swan.

But *I* is not true. My uncle Ernie saw a black swan, and then
took me to see it too.

Therefore, reject H.

A Deductive Argument

The germ theory of disease was discovered through the application
of deduction. Ignaz Semmelweis was a Hungarian physician who
conducted a set of experiments (twelve years before Pasteur's germ
and bacteria research) to determine what was causing high mortality
rates at a maternity ward in the Vienna General Hospital. The scientific method was not well established at that point, but his systematic
observations and manipulations helped not only to pinpoint the cul-
prit, but also to advance scientific knowledge. His experiments are a
model of deductive logic and scientific reasoning.

Built into the scenario was a kind of control condition: The
Vienna General had two maternity wards adjacent to each other, the
first division (with the high mortality rate) and the second division
(with a low mortality rate). No one could figure out why infant and
mother death rates were so much higher in one ward than the other.

One explanation offered by a board of inquiry was that the con-
figuration of the first division promoted psychological distress:
Whenever a priest was called in to give last rites to a dying woman,
he had to pass right by maternity beds in the first division to get to
her; this was preceded by a nurse ringing a bell. The combination
was believed to terrify the women giving birth and therefore make

them more likely victims of this "childbed fever." The priest did not
have to pass by birthing mothers in the second division when he
delivered last rites because he had direct access to the room where
dying women were kept.

Semmelweis proposed a hypothesis and implication that
described an experiment:

> H: The presence of the ringing bell and the priest increases
>   chances of infection.
> I: If the bell and priest are not present, infection is not
>   increased.

Semmelweis persuaded the priest to take an awkward, circuitous
route to avoid passing the birthing mothers of the first division, and
he persuaded the nurse to stop ringing the bell. The mortality rate
did not decrease.

> I is not true.
> Therefore H is false.

We reject the hypothesis after careful experimentation.

Semmelweis entertained other hypotheses. It wasn't overcrowd-
ing, because, in fact, the second division was the more crowded one.
It wasn't temperature or humidity, because they were the same in
the two divisions. As often happens in scientific discovery, a chance
event, purely serendipitous, led to an insight. A good friend of Semmelweis's was accidentally cut by the scalpel of a student who had
just finished performing an autopsy. The friend became very sick,
and the subsequent autopsy revealed some of the same signs of

infection as were found in the women who were dying during child-birth. Semmelweis wondered if there was a connection between the particles or chemicals found in cadavers and the spread of the disease. Another difference between the two divisions that had seemed irrelevant now suddenly seemed relevant: The first division staff were medical students, who were often performing autopsies or cadaver dissections when they were called away to deliver a baby; the second division staff were midwives who had no other duties. It was not common practice for doctors to wash their hands, and so Semmelweis proposed the following:

H: The presence of cadaverous contaminants on the hands of doctors increases chances of infection.
I: If the contaminants are neutralized, infection is not increased.

Of course, an alternative *I* was possible too: If the workers in the two divisions were switched (if midwives delivered in division one and medical students in division two) infection would be decreased. This is a valid implication too, but for two reasons switching the workers was not as good an idea as getting the doctors to wash their hands. First, if the hypothesis was really true, the death rate at the hospital would remain the same—all Semmelweis would have done was to shift the deaths from one division to another. Second, when not delivering babies, the doctors still had to work in their labs in division one, and so there would be an increased delay for both sets of workers to reach mothers in labor, which could contribute to additional deaths. Getting the doctors to wash their hands had the advantage that if it worked, the death rate throughout the hospital would be lowered.

Semmelweis conducted the experiment by asking the doctors to disinfect their hands with a solution containing chlorine. The mortality rate in the first division dropped from 18 percent to under 2 percent.

LOGICAL FALLACIES

Illusory Correlation

The brain is a giant pattern detector, and it seeks to extract order and structure from what often appear to be random configurations. We see Orion the Hunter in the night sky not because the stars were organized that way but because our brains can project patterns onto randomness.

When that friend phones you just as you're thinking of them, that kind of coincidence is so surprising that your brain registers it. What it doesn't do such a good job of is registering all the times you *didn't* think of someone and they called you. You can think of this like one of those fourfold tables from Part One. Suppose it's a particularly amazing week filled with coincidences (a black cat crosses your path as you walk by a junkyard full of broken mirrors, make your way up to the thirteenth floor of a building to find the movie *Friday the 13th* playing on a television set there). Let's say you get twenty phone calls that week and two of them were from long-lost friends whom you hadn't thought about for a while, but they called within ten minutes of you thinking of them. That's the top row of your table: twenty calls, two that you summoned using extrasensory signaling, eighteen that you didn't. But wait! We have to fill in the bottom row of the table: How many times were you thinking about people and they *didn't* call, and—here's my favorite—how many times were you *not* thinking about someone and they didn't call?

Was I Thinking About Them Just Before?

		YES	NO	
Someone Phoned	**YES**	2	18	20
	NO	50	930	980
		52	948	1,000

To fill out the rest of the table, let's say there are 52 times in a week that you're thinking about people, and 930 times in a week when you are not thinking about people. (This last one is just a crazy guess, but if we divide up the 168-hour week into ten-minute increments, that's about 980 total thoughts, and we already know that 50 of those were about people who didn't phone you, leaving 930 thoughts about things other than people; this is probably an underestimate, but the point is made with any reasonable number you care to put here—try it yourself.)

The brain really only notices the upper left-hand square and ignores the other three, much to the detriment of logical thinking (and to the encouragement of magical thinking). Now, before you book a trip to Vegas to play the roulette wheel, let's run the numbers. What is the probability that someone will call *given* that you just thought about them? It's only two out of fifty-two, or 4 percent.

That's right, 4 percent of the time when you think of someone they call you. That's not so impressive.

What might account for the 4 percent of the times when this coincidence occurs? A physicist might just invoke the 1,000 events in your fourfold table and note that only two of them (two-tenths of 1 percent) appear to be "weird" and so you should just expect this by chance. A social psychologist might wonder if there was some external event that caused both you and your friend to think of each other, thus prompting the call. You read about the terrorist attacks in Paris on November 13, 2015. Somewhere in the back of your mind, you remember that you and a college friend always talked about going to Paris. She calls you and you're so surprised to hear from her you forget the Paris connection, but she is reacting to the same event, and that's why she picked up the phone.

If this reminds you of the twins-reared-apart story earlier, it should. Illusory correlation is the standard explanation offered by behavioral geneticists for the strange confluence of behaviors, such as both twins scratching their heads with their middle finger, or both wrapping tape around pens and pencils to improve their grip. We are fascinated by the contents of the upper left-hand cell in the fourfold table, fixated on all the things that the twins do in common. We tend to ignore all the things that one twin does and the other doesn't.

Framing of Probabilities

After that phone call from your old college friend, you decide to go to Paris on vacation for a week next summer. While standing in front of the *Mona Lisa*, you hear a familiar voice and look up to see your old college roommate Justin, whom you haven't seen in years.

"I can't believe it!" Justin says. "I know!" you say. "What are the odds that I'd run into you here in Paris, standing right in front of the *Mona Lisa*! They must be millions to one!"

Yes, the odds of running into Justin in front of the *Mona Lisa* are probably millions to one (they'd be difficult to calculate precisely, yet any calculation you do would make clear that this was very unlikely). But this way of framing the probability is fallacious. Let's take a step back. What if you hadn't run into Justin just as you were standing in front of the *Mona Lisa*, but as you were in front of the *Venus de Milo*, in *les toilettes*, or even as you were walking in the entrance? What if you had run into Justin at your hotel, at a café, or the Eiffel Tower? You would have been just as surprised. For that matter, forget about Justin—if you had run into *anyone you knew* during that vacation, *anywhere in Paris,* you'd be just as surprised. And why limit it to your vacation in Paris? It could be on a business trip to Madrid, while changing planes in Cleveland, or at a spa in Tucson. Let's frame the probability this way: Sometime in your adult life, you'll run into someone you know where you wouldn't expect to run into them. Clearly the odds of that happening are quite good. But the brain doesn't automatically think this way— cognitive science has shown us just how necessary it is for us to train ourselves to avoid squishy thinking.

Framing Risk

A related problem in framing probabilities is the failure to frame risks logically. Even counting the airplane fatalities of the 9/11 attacks in the United States, air travel remained (and continues to remain) the safest transportation mode, followed closely by rail transportation.

The chances of dying on a commercial flight or train trip are next to zero. Yet, right after 9/11, many U.S. travelers avoided airplanes and took to the highways instead. Automobile deaths increased dramatically. People followed their emotional intuition rather than a logical response, oblivious to the increased risk. The *rate* of vehicular accidents did not increase beyond baseline, but the sum of people who died in all transportation-related accidents increased as more people chose a less safe mode of travel.

You might pull up a statistic such as this one:

More people died in plane crashes in 2014 than in 1960.

From this, you might conclude that air travel has become much less safe. The statistic is correct, but it's not the statistic that's relevant. If you're trying to figure out how safe air travel is, looking at the total number of deaths doesn't tell you that. You need to look at the death *rate*—the deaths per miles flown, or deaths per flight, or something that equalizes the baseline. There were not nearly as many flights in 1960, but they were more dangerous.

By similar logic, you can say that more people are killed on highways between five and seven p.m. than between two and four a.m., so you should avoid driving between five and seven. But the simple fact is that many times more people are driving between five and seven—you need to look at the *rate* of death (per mile or per trip or per car), not the raw number. If you do, you'll find that driving in the evening is safer (in part because people on the road between two and four a.m. are more likely to be drunk or sleep-deprived).

After the Paris attacks of November 13, 2015, CNN reported that at least one of the attackers had entered the European Union as a refugee, against a backdrop of growing anti-refugee sentiment in Europe. Anti-refugee activists had been calling for stricter border control. This is a social and political issue and it is not my intention to take a stand on it, but the numbers can inform the decision making. Closing the borders completely to migrants and refugees might have thwarted the attacks, which took roughly 130 lives. Denying entry to a million migrants coming from war-torn regions such as Syria and Afghanistan would, with great certainty, have cost thousands of them their lives, far more than the 130 who died in the attacks. There are other risks to both courses of action, and other considerations. But to someone who isn't thinking through the logic of the numbers, a headline like "One of the attackers was a refugee" inflames the emotions around anti-immigrant sentiment, without acknowledging the many lives that immigration policies saved. The lie that terrorists want you to believe is that you are in immediate and great peril.

Misframing is often used by salespeople to persuade you to buy their products. Suppose you get an email from a home-security company with this pitch: "Ninety percent of home robberies are solved with video provided by the homeowner." It sounds so empirical. So scientific.

Start with a plausibility check. Forget about the second part of the sentence, about the video, and just look at the first part: "Ninety percent of home robberies are solved . . ." Does that seem reasonable? Without looking up the actual statistics, just using your real-world knowledge, it seems doubtful that 90 percent of home robberies are solved. This would be a fantastic success rate for any police department. Off to the Internet. An FBI page reports that about 30 percent of robbery cases are "cleared," meaning solved.

So we can reject as highly unlikely the initial statement. It said 90 percent of home robberies are solved with video provided by the homeowner. But that can't be true—it would imply that more than 90 percent of home robberies are solved, because some are certainly solved without home video. What the company more likely means is that 90 percent of solved robberies are from video provided by the homeowner.

Isn't that the same thing?

No, because the sample pool is different. In the first case, we're looking at all home robberies committed. In the second case we're looking only at the ones that were solved, a much smaller number. Here it is visually:

All home robberies in a neighborhood:

**Solved home robberies in a neighborhood
(using the 30 percent figure obtained earlier):**

So does that mean that if I have a video camera there is a 90 percent chance that the police will be able to solve a burglary at my house?

No!

All you know is that *if* a robbery is solved, there is a 90 percent chance that the police were aided by a home video. If you're thinking that we have enough information to answer the question you're really interested in (what is the chance that the police will be able to solve a burglary at my house if I buy a home-security system versus if I don't), you're wrong—we need to set up a fourfold table like the ones in Part One, but if you start, you'll see that we have information on only one of the rows. We know which percentage of solved crimes had home video. But to fill out the fourfold table, we'd also need to know what proportion of the unsolved crimes had home video (or, alternatively, what proportion of the home videos taken resulted in unsolved crimes).

Remember, P(burglary solved | home video) ≠ P(home video | burglary solved).

The misframing of the data is meant to spark your emotions and cause you to purchase a product that may not have the intended result at all.

Belief Perseverance

An odd feature of human cognition is that once we form a belief or accept a claim, it's very hard for us to let go, even in the face of overwhelming evidence and scientific proof to the contrary. Research reports say we should eat a low-fat, high-carb diet and so we do.

New research undermines the earlier finding—quite convincingly—yet we are reluctant to change our eating habits. Why? Because on acquiring the new information, we tend to build up internal stories to help us assimilate the knowledge. "Eating fats will make me fat," we tell ourselves, "so the low-fat diet makes a lot of sense." We read about a man convicted of a grisly homicide. We see his picture in the newspaper and think that we can make out the beady eyes and unforgiving jaw of a cold-blooded murderer. We convince ourselves that he "looks like a killer." His eyebrows arched, his mouth relaxed, he seems to lack remorse. When he's later acquitted based on exculpatory evidence, we can't shake the feeling that even if he didn't do *this* murder, he must have done another one. Otherwise he wouldn't look so guilty.

In a famous psychology experiment, participants were shown photos of members of the opposite sex while ostensibly connected to physiological monitoring equipment indicating their arousal levels. In fact, they weren't connected to the equipment at all—it was under the experimenter's control. The experimenter gave the participants false feedback to make them believe that they were particularly attracted to a person in one of the photos more than the others. When the experiment was over, the participants were shown that the "reactions" of their own body were in fact premanufactured tape recordings. The kicker is that the experimenter allowed them to choose one photo to take home with them. Logically, they should have chosen the picture that they found most attractive at that moment—the evidence for liking a particular picture had been completely discredited. But the participants tended to choose the picture that was consistent with their initial belief. The experimenters showed that the effect was driven by the sort of self-persuasion described above.

Autism and Vaccines: Four Pitfalls in Reasoning

The story with autism and vaccines involves four different pitfalls in critical thinking: illusory correlation, belief perseverance, persuasion by association, and the logical fallacy we saw earlier, *post hoc, ergo propter hoc* (loosely translated, it means "because this happened after that, that must have caused this").

Between 1990 and 2010, the number of children diagnosed with autism spectrum disorders (ASD) rose sixfold, more than doubling

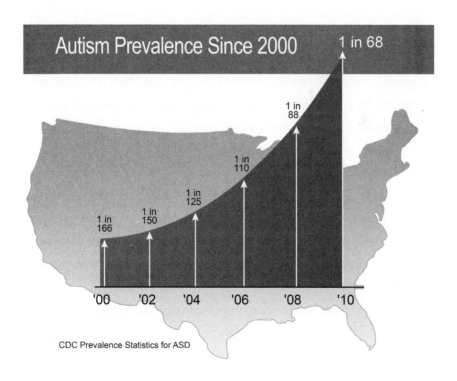

Autism Prevalence Since 2000 — 1 in 68

1 in 88

1 in 110

1 in 125

1 in 150

1 in 166

'00 '02 '04 '06 '08 '10

CDC Prevalence Statistics for ASD

in the last ten years. The prevalence of autism has increased exponentially from the 1970s to now.

The majority of the rise has been accounted for by three factors: increased awareness of autism (more parents are on the alert and bring their children in for evaluation, professionals are more willing to make the diagnosis); widened definitions that include more cases; and the fact that people are having children later in life (advanced parental age is correlated with the likelihood of having children with autism and many other disorders).

If you allow the Internet to guide your thinking on why autism has increased, you'll be introduced to a world of fiendish culprits: GMOs, refined sugar, childhood vaccines, glyphosates, Wi-Fi, and proximity to freeways. What's a concerned citizen to do? It sure would be nice if an expert would weigh in. Voilà—an MIT scientist comes to the rescue! Dr. Stephanie Seneff made headlines in 2015 when she reported a link between a rise in the use of glyphosate, the active ingredient in the weed killer Roundup, and the rise of autism. That's right, two things rise—like pirates and global warming—so there must be a causal connection, right?

Post hoc, ergo propter hoc, anyone?

Dr. Seneff is a computer scientist with no training in agriculture, genetics, or epidemiology. But she is a *scientist* at the venerable MIT, so many people wrongly assume that her expertise extends beyond her training. She also couches her argument in the language of science, giving it a real pseudoscientific, counterknowledge gloss:

1. Glyphosate interrupts the shikimate pathway in plants.
2. The shikimate pathway allows plants to create amino acids.
3. When the pathway is interrupted, the plants die.

Seneff concedes that human cells don't have a shikimate pathway, but she continues:

4. We have millions of bacteria in our gut ("gut flora").
5. Those bacteria do have a shikimate pathway.
6. When glyphosate enters our system, it disturbs our digestion and our immune function.
7. Glyphosate in humans can also inhibit liver function.

If you're wondering what all this has to do with ASD, you should be. Seneff lays out a case (without citing any evidence) for increased prevalence of digestive problems and immune system dysfunction, but these have nothing to do with ASD.

Others searching for an explanation for the rise in autism rates have pointed to the MMR (measles-mumps-rubella) vaccine, and the antiseptic, antifungal compound thimerosal (thiomersal) it contains. Thimerosal is derived from mercury, and the amount contained in vaccines is typically one-fortieth of what the World Health Organization (WHO) considers the amount tolerable per day. Note that the WHO guidelines are expressed as per-day amounts, and with the vaccine, you're getting it only once.

Although there was no evidence that thimerosal was linked to autism, it was removed from vaccines in 1992 in Denmark and Sweden, and in the United States starting in 1999, as a "precautionary measure." Autism rates have continued to increase apace even with the agent removed. The illusory correlation (as in pirates and global warming) is that the MMR vaccine is typically given between twelve and fifteen months of age, and if a child has autism, the earliest that it is typically diagnosed is between eighteen and twenty-four months of age. Parents tended to focus on the upper left-hand cell of a fourfold table—the number of times a child received a vaccination and was later diagnosed with autism—without considering how many children who were not vaccinated still developed autism, or how many *millions* of children were vaccinated and did not develop autism.

To make matters worse, a now-discredited physician, Andrew Wakefield, published a scientific paper in 1998 claiming a link. The *British Medical Journal* declared his work fraudulent, and six years later, the journal that originally published it, the *Lancet*, retracted it. His medical license was revoked. Wakefield was a surgeon, not an expert in epidemiology, toxicology, genetics, neurology, or any specialization that would have qualified him as an expert on autism.

Post hoc, ergo propter hoc caused people to believe the correlation implied causation. Illusory correlation caused them to focus only on the coincidence of some people developing autism who also had the vaccine. The testimony of a computer scientist and a physician caused people to be persuaded by association. Belief perseverance caused people who initially believed the link to cling to their beliefs even after the evidence had been removed.

Parents continue to blame the vaccine for autism, and many parents stopped vaccinating their children. This led to several outbreaks of measles around the world. All because of a spurious link and the failure of a great many people to distinguish between correlation and causation, and a failure to form beliefs based on what is now overwhelming scientific evidence.

KNOWING WHAT YOU DON'T KNOW

. . . as we know, there are known knowns; there are things
we know we know. We also know there are known
unknowns; that is to say we know there are some things we
do not know. But there are also unknown unknowns—the
ones we don't know we don't know.

—U.S. Secretary of Defense Donald Rumsfeld

This is clearly tortured language, and the meaning of the sentence
is obscured by that. There's no reason for the repetitive use of the
same word, and the secretary might have been clearer if he had said
instead, "There are things we know, things we are aware that we do
not know, and some things we aren't even aware that we don't
know." There's a fourth possibility, of course—things we know that
we aren't aware we know. You've probably experienced this—
someone asks you a question and you answer it, and then say to
yourself, "I'm not even sure how I knew that."

Either way, the fundamental point is sound, you know? What
will really hurt you, and cause untold amounts of damage and
inconvenience, are the things you think you know but don't (per
Mark Twain's/Josh Billings's epigraph at the beginning of this
book), and the things that you weren't even aware of that are

supremely relevant to the decision you have ahead (the unknown unknowns). Formulating a proper scientific question requires taking an account of what we know and what we don't know. A properly formulated scientific hypothesis is *falsifiable*—there are steps we can take, at least in theory, to test the true state of the world, to determine if our hypothesis is true or not. In practice, this means considering alternative explanations ahead of time, before conducting the experiment, and designing the experiment so that the alternatives are ruled out.

If you're trying out a new medicine on two groups of people, the experimental conditions have to be the same in order to conclude that medicine A is better than medicine B. If all the people in group A get to take their medicine in a windowed room with a nice view, and the people in group B have to take it in a smelly basement lab, you've got a confounding factor that doesn't allow you to conclude the difference (if you find one) was due solely to the medication. The smelly basement problem is a known known. Whether medicine A works better than medicine B is a known unknown (it's why we're conducting the experiment). The unknown unknown here would be some other potentially confounding factor. Maybe people with high blood pressure respond better to medicine A in every case, and people with low blood pressure respond better to medicine B. Maybe family history matters. Maybe the time of day the medication is taken makes a difference. Once you identify a potential confounding factor, it very neatly moves from the category of unknown unknown to known unknown. Then we can modify the experiment, or do additional research that will help us to find out.

The trick to designing good experiments—or evaluating ones that have already been conducted—comes down to being able to generate alternative explanations. Uncovering unknown unknowns might be said to be *the* principal job of scientists. When experiments yield surprising results, we rejoice because this is a chance to learn something we didn't know. The B-movie characterization of the scientist who clings to his pet theory to his last breath doesn't apply to any scientist I know; real scientists know that they only learn when things don't turn out the way they thought they would.

In a nutshell:

1. There are some things we know, such as the distance from the Earth to the sun. You may not be able to generate an answer without looking it up, but you are aware that the answer is known. This is Rummy's *known known*.

2. There are some things that we don't know, such as how neural firing leads to feelings of joy. We're aware that we don't know the answer to this. This is Rummy's *known unknown*.

3. There are some things that we know, but we aren't aware that we know them, or forget that we know them. What is your grandmother's maiden name? Who sat next to you in third grade? If the right retrieval cues help you to recollect something, you find that you knew it, although you didn't realize ahead of time that you did. Although Rumsfeld doesn't mention them, this is an *unknown known*.

4. There are some things that we don't know, and we're not even aware we don't know them. If you've bought a house,

you've probably hired various inspectors to report on the condition of the roof, the foundation, and the existence of termites or other wood-destroying organisms. If you had never heard of radon, and your real estate agent was more interested in closing the deal than protecting your family's health, you wouldn't think to test for it. But many homes do have high levels of radon, a known carcinogen. This would count as an *unknown unknown* (although, having read this paragraph, it is no longer one). Note that whether you're aware or unaware of an unknown depends on your expertise and experience. A pest-control inspector would tell you that he is only reporting on what's visible—it is known to him that there might be hidden damage to your house, in areas he was unable to access. The nature and extent of this damage, if any, is unknown to him, but he's aware that it might be there (a *known unknown*). If you blindly accept his report and assume it is complete, then you're unaware that additional damage could exist (an *unknown unknown*).

We can clarify Secretary Rumsfeld's four possibilities with a four-fold table:

What we know that we know: GOOD—PUT IT IN THE BANK	What we know that we don't know: NOT BAD, WE CAN LEARN IT
What we don't know that we know: A BONUS	What we don't know that we don't know: DANGER—HIDDEN SHOALS

The unknown unknowns are the most dangerous. Some of the biggest human-caused disasters can be traced to these. When bridges collapse, countries lose wars, or home purchasers face foreclosure, it's often because someone didn't allow for the possibility that they don't know everything, and they proceeded along blindly thinking that every contingency had been accounted for. One of the main purposes of training someone for a PhD, a law or medical degree, an MBA, or military leadership is to teach them to identify and think systematically about what they don't know, to turn unknown unknowns into known unknowns.

A final class that Secretary Rumsfeld didn't talk about either are incorrect knowns—things that we think are so, but aren't. Believing false claims falls into this category. One of the biggest causes of bad, even fatal, outcomes is belief in things that are untrue.

BAYESIAN THINKING IN SCIENCE AND IN COURT

Recall from Part One the idea of Bayesian probability, in which you can modify or update your belief about something based on new data as it comes in, or on the prior probability of something being true—the probability you have pneumonia *given* that you show certain symptoms, or the probability that a person will vote for a particular party *given* where they live.

In the Bayesian approach, we assign a subjective probability to the hypothesis (the *prior* probability), and then modify that probability in light of the data collected (the *posterior* probability, because it's the one you arrive at after you've conducted the experiment). If we had reason to believe the hypothesis was true before we tested it, it doesn't take much evidence for us to confirm it. If we had reason to believe the hypothesis unlikely before we tested it, we need more evidence.

Unlikely claims, then, according to a Bayesian perspective, require stronger proof than likely ones. Suppose your friend says she saw something flying right outside the window. You might entertain three hypotheses, *given* your own recent experiences at that window: It is a robin, it is a sparrow, or it is a pig. You can assign probabilities to these three hypotheses. Now your friend shows you a photo of a pig flying outside the window. Your prior belief that pigs fly is so small that the posterior probability is still very small, even with this evidence. You're probably now entertaining new hypotheses that the photo was doc-tored, or that there was some other kind of trickery involved. If this reminds you of the fourfold tables and the likelihood that someone has breast cancer given a positive test, it should—the fourfold tables are simply a method for performing Bayesian calculations.

Scientists should set a higher threshold for evidence that goes against standard theories or models than for evidence that is con-sistent with what we know. Following thousands of successful trials for a new retroviral drug in mice and monkeys, when we find that it works in humans we are not surprised—we're willing to accept the evidence following standard conventions for proof. We might be convinced by a single study with only a few hundred partici-pants. But if someone tells us that sitting under a pyramid for three days will cure AIDS, by channeling qi into your chakras, this requires stronger evidence than a single experiment because it is farfetched and nothing like it has ever been demonstrated before. We'd want to see the result replicated many times and under many different conditions, and ultimately, a meta-analysis.

The Bayesian approach isn't the only way that scientists deal with unlikely events. In their search for the Higgs boson, physicists set a threshold (using conventional, not Bayesian, statistical tests) 50,000 times more stringent than usual—not because the Higgs was unlikely (its existence was hypothesized for decades) but because the cost of being wrong is very high (the experiments are very expensive to conduct).

The application of Bayes's rule can perhaps best be illustrated with an example from forensic science. One of the cornerstone prin-ciples of forensic science was developed by the French physician and lawyer Edmond Locard: Every contact leaves a trace. Locard stated that either the wrongdoer leaves signs at the scene of the crime or has taken away with her—on her person, body, or clothes—indications of where she has been or what she has done.

Suppose a criminal breaks into the stables to drug a horse the night before a big race. He will leave some traces of his presence at the crime scene—footprints, perhaps skin, hair, clothing fibers, etc. Evidence has been transferred from the criminal to the scene of the crime. And similarly, he will pick up dirt, horsehair, blanket fibers, and such from the stable, and in this way evidence has been transferred from the crime scene to the criminal.

Now suppose someone is arrested the next day. Samples are taken from his clothing, hands, and fingernails, and similarities are found between these samples and other samples taken at the crime scene. The district attorney wants to evaluate the strength of this evidence. The similarities may exist because the suspect is guilty. Or perhaps the suspect is innocent, but was in contact with the guilty party—that contact too would leave a trace. Or perhaps the suspect, quite innocently, was in another barn, interacting innocently with another horse, accounting for the similarities.

Using Bayes's rule allows us to combine objective probabilities, such as the probability of the suspect's DNA matching the DNA found at the crime scene, with personal, subjective views, such as the credibility of a witness, or the honesty and track record of the CSI officer who had custody of the DNA sample. Is the suspect someone who has done this before, or someone who knows nothing about horse racing, has no connection to anyone involved in the race, and has a very good alibi? These factors help us to determine a prior, subjective probability that the suspect is guilty.

If we take literally the assumption in the American legal system that one is innocent until proven guilty, then the prior probability of a suspect being guilty is zero, and any evidence, no matter how damning, won't yield a posterior probability above zero, because you'll always be multiplying by zero. A more reasonable way to establish the prior probability of a suspect's innocence is to consider anyone in the population equally likely. Thus, if the suspect was apprehended in a city of 100,000 people, and investigators have reason to believe that the perpetrator was a resident of the city, the prior odds of the suspect being guilty are 1 in 100,000. Of course, evidence can narrow the population—we may know, for example, that there were no signs of forced entry, and so the suspect had to be one of fifty people who had access to the facility.

Our prior hypothesis (*a priori* in Latin) is that the suspect is guilty with a probability of .02 (one of fifty people who had access). Now let's suppose the perpetrator and the horse got in a scuffle, and human blood was found at the scene. Our forensics team tells us that the probability that the suspect's blood matches the blood found at the scene is .85. We construct a fourfold table as before. We fill in the bottom row under the table first: The suspect has a one in fifty chance to be guilty (the *Guilty: Yes* column), and a forty-nine in fifty chance to be innocent. The lab told us that there's a .85 probability of a blood match, so we enter that in the upper left: the probability that the suspect is guilty *and* the blood matches. That means the lower left cell has to be .15 (the probabilities have to add up to one). The .85 blood match means something else: that there's a .15 chance the blood was left by someone else, not our suspect, which would absolve him and render him not guilty. There's a .15 chance that one of the people in the right-hand column will match, so we multiply 49 × .15 to get 7.35 in the upper right cell. We subtract that from the forty-nine in order to find the value for the bottom right cell.

Suspect Guilty

		YES	NO	
Blood Match	**YES**	0.85	7.35	8.2
	NO	0.15	41.65	41.8
		1	49	50

Now we can calculate the information we want the judge and jury to evaluate.

$$P(Guilty \mid Match) = .85/8.2 = .10$$
$$P(Innocent \mid Match) = 7.35/8.2 = .90$$

Given the evidence, it is about nine times more likely that our suspect is innocent than guilty. We started out with him having a .02 chance of being guilty, so the new information has increased his guilt by a factor of five, but it is still more likely that he is innocent.

Suppose, however, some new evidence comes in—horsehair found on the suspect's coat—and the probability that the horsehair belongs to the drugged horse is .95 (only five chances in one hundred that the hair belongs to a different horse). We can chain our Bayesian probabilities together now, filling out a new table. In the bottom margin, we enter the values we just calculated, .10 and .90. (Statisticians sometimes say that yesterday's posteriors are today's priors.) If you'd rather think of these numbers as "one chance in ten" and "nine chances in ten," go ahead and enter them as whole numbers.

Suspect Guilty

		YES	NO	
Blood Match	**YES**	0.95	0.45	1.4
	NO	0.05	8.55	8.6
		1	9	10

We know from our forensics team that the probability of a match for the hair sample is .95. Multiplying that by one, we get the entry for the upper left, and subtracting that from one we get the entry for the lower left. If there is a .95 chance that the sample matches the victimized horse, that implies that there is a .05 chance that the sample matches a different animal (which would absolve the suspect) so the upper right-hand cell is the product of .05 and the marginal total of 9 = .45. Now when we perform our calculations, we see that

P(Guilty | Evidence) = .68   P(Evidence | Guilty) = .95

P(Innocent | Evidence) = .32   P(Evidence | Innocent) = .05

The new evidence shows us that it is about twice as likely that the suspect is guilty as that he is innocent, given the evidence. Many attorneys and judges do not know how to organize the evidence like this, but you can see how helpful it is. The problem of mistakenly thinking that P(Guilty | Evidence) = P(Evidence | Guilt) is so widespread it has been dubbed *the prosecutor's fallacy.*

If you prefer, the application of Bayes's rule can be done mathematically, rather than using the fourfold table, and this is shown in the appendix.

Four Case Studies

Science doesn't present us with certainty, only probabilities. We don't know for 100 percent sure that the sun will come up tomorrow, or that the magnet we pick up will attract steel, or that nothing travels faster than the speed of light. We think these things very likely, but science yields only the best Bayesian conclusions we can have, given what we know so far.

Bayesian reasoning asks us to consider probabilities in light of what we know about the state of the world. Crucial to this is engaging in the kind of critical thinking described in this field guide. Critical thinking is something that can be taught, and practiced, and honed as a skill. Rigorous study of particular cases is a standard approach because it allows us to practice what we've learned in new contexts—what learning theorists call *far transfer*. Far transfer is the most effective way we know to make knowledge stick.

There is an infinite variety of ways that faulty reasoning and misinformation can sneak up on us. Our brains weren't built to excel at this. It's always been a part of science to take a step back and engage in careful, systematic reasoning. Case studies are presented as stories, based on true incidents or composites of true incidents, and of course, we are a story-loving species. We remember the stories and the interesting way they loop back to the fundamental concepts. Think of the following as problem sets we can all explore together.

Shadow the Wonder Dog Has Cancer (or Does He?)

We got our dog Shadow, a Pomeranian-Sheltie mix, from a rescue shelter when he was two years old. He got his name, we learned, because he would follow us from room to room around the house during the day, never far away. As often happens with pets, our rhythms synchronized—we would fall asleep and wake up at the same time, get hungry around the same time, feel like getting exercise at the same time. He traveled with us often on business trips to other cities, becoming acclimated to planes, trains, and automobiles.

When Shadow was thirteen, he began having trouble urinating, and one morning we found blood in his urine. Our vet conducted an ultrasound examination and found a growth on his bladder. The only way to tell whether it was cancerous was to perform two surgical procedures that the oncologist was urging: a cystoscopy, which would run a miniature camera through his urethra into the bladder, and a biopsy to sample the mass and study it under the microscope. The general practitioner cautioned against this because of the risks of general anesthesia in a dog Shadow's age. If it did turn out to be cancerous, the oncologist would want to perform surgery and start chemotherapy. Without any further tests, the doctors were still pretty certain that this was bladder cancer, known as transitional cell carcinoma (TCC). On average, dogs live only six months following this diagnosis.

As my wife and I looked into Shadow's eyes, we felt utterly helpless. We didn't know if he was in pain, and if so, how much more he was facing, either from the treatment or from the disease. His care was entirely in our hands. This made the decision particularly emotional, but that didn't mean we threw rationality out the window. You can think critically even when the decision is emotional. Even when it's your dog.

This is a typical medical scenario for people or pets: two doctors, two different opinions, many questions. What are the risks of surgery? What are the risks of the biopsy? How long is Shadow likely to live if we give him the operation and how long is he likely to live if we don't?

In a biopsy, a small needle is used to collect a sample of tissue that is then sent to a pathologist, who reports on the likelihood that it is cancerous or not. (Pathology, like most science we've seen, does not deal in certainties, just likelihoods and the probability that the sample contains cancer, which is then applied to the probability that the unsampled parts of the organ might also contain cancer; if you're looking for certainty, pathology is not the place to look.) Patients and pet owners almost never ask about the risk of biopsy. For humans, these statistics are well known, but they are less well tracked in veterinary medicine. Our vet estimated that there was a 5 percent chance of life-threatening infection, and a 10 percent chance that some cancerous material (if indeed the mass was cancerous) would be "shed" into the abdomen on the needle's way out, seeding further cancer growth. An additional risk was that biopsies leave behind scar tissue that makes it more difficult to operate later if that's what you decide to do. The anesthesia needed for the procedure could kill Shadow. In short, the diagnostic procedure could make him worse.

Our vet presented us with six options:

1. Biopsy through the abdominal wall in the hope of obtaining a more definitive diagnosis.
2. Diagnostic catheterization (using a catheter to traumatize a portion of the mass, allowing cells to exfoliate and then be examined).
3. Biopsy using the same cystoscopic camera they wanted to use anyway to better image the mass (through the urethra).
4. Major surgery right now to view the mass directly, and remove it if possible. The problem with this is that most bladder cancers return within twelve months because the surgeons are unable to remove every cancerous cell, and the ones left behind typically keep on growing at a rapid rate.
5. Do nothing.
6. Put Shadow to sleep right now, in recognition of the fact that it most probably is bladder cancer, and he doesn't have long to live anyway.

We asked about what the treatment options were if it was found to be cancer, and what they might be if it was not cancer. Too often, patients focus on the immediate, upcoming procedure without regard for what the next steps might be.

If the mass was cancerous, the big worry was that the tumor could grow and eventually block one of the tubes that brings urine into the bladder from the kidneys, or that allows urine to leave the bladder and end up on a lawn or fire hydrant of choice. If that blockage occurs, Shadow could experience great pain and die within a day. Along the way to this, there could be temporary blockages as a result of swelling. Because of the position of the bladder within the body, and the angle of ultrasound, it was difficult to tell how close the mass was to these tubes (the ureter and urethra).

So what about the six options presented above—how to decide which (if any) to choose? We ruled out two of them: putting Shadow to sleep and doing nothing. Recall that the oncologist was pushing for surgery because that is their gold standard, their protocol for such cases. We asked for some statistics and she said she'd have to do some research and get back to us. Later, she said that there was a 20 percent chance that the surgery would end badly, killing Shadow right away. So we ruled out the major surgery because we weren't even sure yet if the mass was cancerous.

We asked for life-expectancy statistics on the various remaining scenarios. Unfortunately, most such statistics are not kept by the veterinary community, and in any case, those that are kept skew toward short life expectancy because many pet owners choose euthanasia. That is, many owners opt to put their pets down before the disease progresses because of concerns about either the animal's quality of life or the owners' quality of life: Dogs with TCC often experience incontinence (we had already noticed that Shadow was leaving us little surprises around the house). We didn't have a definitive diagnosis yet, but based on the sparse statistics that existed, it looked as though Shadow would live three months *with or without treatment*. Three months if we do nothing, three months if we give him chemo, three months if we give him surgery. How could that be? Ten years ago, we found out, vets would recommend euthanasia on first diagnosis of TCC. And at the first sign of chronic incontinence, owners would put their dogs down. So owners were typically ending their dogs' lives before the cancer did, and this made the statistics unreliable.

We did some research on our own, using "transitional cell

carcinoma" and "dog *or* canine" as the search terms. We found out that there was a 30 percent chance Shadow could improve simply by taking a nonsteroidal anti-inflammatory called Piroxicam. Piroxicam has its own side effects, including stomach upset, vomiting, loss of appetite, and kidney and liver trouble. We asked the vet about it and she agreed that it made sense to start him on Piroxicam no matter what else we were doing.

From the Purdue University website—Purdue has one of the leading veterinary medical centers—we were able to obtain the following survival statistics:

1. Median survival with major surgery = 109 days
2. Median survival with chemotherapy = 130 days
3. Median survival with Piroxicam = 195 days

The range of survival times in all of these studies, however, varied tremendously from dog to dog. Some dogs died after only a few days, while others lived more than two years.

We decided that the most rational choice was to start Shadow on the Piroxicam because its side effects were relatively minor, compared to the others, and to get the cystoscopy in order to give the doctor a better look at the mass and the associated biopsy to give us more to go on. Shadow would have to be lightly anesthetized, but it was only for a short time and the doctors were confident that he would emerge fine.

Two weeks later, the cystoscopy showed that the mass was in fact very close to the ureters and the urethral openings—so close, in fact, that surgery wouldn't help if the mass was cancerous because too much of the tumor would be left behind. The pathologist wasn't able to tell if the tissue was cancerous or not because the procedure ended up not getting a large enough sample. So after all that, we still didn't have a diagnosis. Yet the statistics above suggested that if Shadow was among the 30 percent of dogs for whom Piroxicam worked, that would yield the best life expectancy. We wouldn't have to subject him to the discomforts of surgery or chemo, and we could just enjoy our time together at home.

There are many instances, with both pets and humans, that a treatment doesn't statistically improve your life expectancy. Taking a statin if you are not in a high-risk group or surgically removing the prostate for cancer if you do not have fast-moving prostate cancer are both treatments with negligible impact on life expectancy. It sounds counterintuitive, but it's true: Not all treatments actually help. It's clear that Shadow would be better off without the surgery (so that we could avoid the 20 percent chance it would kill him) and the chemo wouldn't buy him any time, statistically.

Shadow responded to the Piroxicam very well and within three days he was back to himself—energetic, in a good mood, happy. By one week he had no more difficulty urinating. We saw occasional minor amounts of blood in the urine, but we were told this was normal after biopsy. Then, 161 days after the initial suspicion (which was never confirmed) of TCC, his kidneys started to fail. We checked him into a specialty oncology clinic. The doctors weren't sure whether the organ failure might be related to TCC, or why it was occurring now. They prescribed medications to address common kidney conditions and ran dozens of tests without getting any closer to understanding what was happening. Shadow grew increasingly uncomfortable and stopped eating. We put him on an IV drip painkiller and two days later, when we took him off for just a few minutes to see how he was doing, he was clearly in pain. We talked to his current and former doctors, carefully describing the situation, its progression, and his condition. All agreed it was time to let him go. We had Shadow's company—and he had ours—for a month longer than the average chemotherapy patient, and during that month he was able to avoid hospitals, catheters, IV lines, and scalpels.

We went to the oncology hospital—the staff knew us well because we had been visiting Shadow there every day in between his tests and treatments—and arranged for him to be put to sleep. He was in pain, and we felt that we had perhaps waited one or two days too long. It was awful to see that large personality suddenly drift away and disappear. We found comfort knowing that we had considered every stage of his care and that he had as good a life as we were able to give him for as long as possible. Perhaps the most difficult emotion that people experience after a disease ends a life is regret over the choices made. We were able to say good-bye to Shadow with no regrets over our decisions. We let our critical thinking, our use of Bayesian reasoning, guide us.

Were Neil Armstrong and Buzz Aldrin Thespians?

Moon-landing deniers point to a number of inconsistencies and unanswered questions. "There should have been more than a two-second delay in communications between the Earth and the moon, because of its distance." "The quality of the photographs is implausibly high." "There are no stars in the sky in any of the photos." "How could the photos of the American flag show ripples in it, as though waving in the air, if the moon has no atmosphere?" The capper is a report by an aerospace worker, Bill Kaysing, who wrote that the probability of a successful landing on the moon was .0017 percent (note the precision of this estimate!). Many more such claims

exist. Part of what keeps counterknowledge going is the sheer number of unanswered questions that keep popping up, like a game of
Whac-A-Mole. If you want to convince people of something that's
not true, it's apparently very effective to simply snow them with one
question after another, and hope that they will be sufficiently
impressed—and overwhelmed—that they won't bother to look for
explanations. But even 1,000 unanswered questions don't necessarily mean that something didn't happen, as any investigator knows.
The websites dedicated to the moon landing denial don't cite the
evidence for it, nor do they publish rebuttals to their claims.

In the case of the moon landing, each of these (and the other
claims) is easily refuted. There *was* a two-second delay in Earth-moon
communications that can be easily heard on the original tapes, but
some documentary films and news reports edited out the delay in the
interest of presenting a more compelling broadcast. The quality of the
photographs is high because the astronauts used a high-resolution
Hasselblad camera with 70mm high-resolution film. There are no
stars in the lunar sky because most of the images we saw were taken
during lunar daytime (otherwise, we wouldn't have been able to see
the astronauts). The flag doesn't show ripples: Aware that there was
no atmosphere, NASA prepared the flag with a t-bar to support its top
edge and the "ripples" are simply folds in the fabric. With no wind to
blow the flag, its creases stay in place. This claim is based on still
photos in which there appears to be a rippling effect, but moving film
images show that the flag is not blowing, it's static.

But what about the report of an aerospace worker that a moon
landing was highly improbable? First, the "aerospace worker" was
not trained in engineering or science; he was a writer with a BA in
English who happened to work for Rocketdyne. The source of his

estimate appears to be from a Rocketdyne report from the 1950s, back when space technology was still in its infancy. Although there are still unanswered questions (e.g., why are some of the original telemetry recordings missing?), the weight of evidence overwhelmingly points to the moon landing being real. It's not certainty, it's just very, very likely. If you're going to use spuriously obtained probability estimates to claim that past events didn't happen, you'd have to similarly conclude that human beings don't really exist: It's been claimed that the chances of life forming on Earth is many billions to one. Like many examples of counterknowledge, this uses the language of science—in this case probability—in a way that utterly debases that fine language.

Statistics Onstage (and in a Box)

David Blaine is a celebrity magician and illusionist. He also claims to have completed great feats of physical endurance (at least one was recognized by the *Guinness Book of World Records*). The question for a critical thinker is: Did he actually demonstrate physical endurance or was he using a clever illusion? Certainly, as a skilled magician, it would be easy for him to fake the endurance work.

In a TED talk with more than 10 million views, he claims to have held his breath for seventeen minutes underwater, and tells us how he trained himself to do it. Other claims are that he froze himself in a block of ice for a week, fasted in a glass box for forty-four days, and was buried alive in a coffin for a week. Are these claims true? Are they even plausible? Are there alternative explanations?

In his videos, Blaine has a down-to-earth manner; he doesn't speak quickly, he doesn't seem slick. He's believable because his speech sounds so awkward that it's difficult to imagine he's calculated just what to say and how to say it. But bear in mind: Professional magicians typically calculate and plan everything they say. Every single move, every apparently spontaneous scratch of the head, is typically rehearsed over and over again. The illusion they're trying to create—the feat of magic—works because the magician is expert at misdirecting your attention and subverting your assumptions about what's spontaneous and what's not.

So how do we apply critical thinking to his endurance performances?

If you're thinking about hierarchies of source quality, you'll focus on the fact that he has a TED talk and TED talks are fact-checked and very tightly curated. Or are they? Well, actually, there are more than 5,000 TED-branded events, but only two are vetted—TED and TEDGlobal. Blaine's video comes from a talk he delivered at TEDMED, one of the more than 4,998 conferences that are run by enthusiasts and volunteers and are not vetted by the TED organization. This doesn't mean it's not true, just that we can't rely on the reputation and authority of TED to establish its truth. Recall TMZ and the reporting of Michael Jackson's death—they're going to be right some of the time, and maybe even a lot of the time, but you can't know for sure.

Before looking at the underwater breath holding, let's look more closely at a couple of Blaine's other claims. For starters, Fox television reported his ice-block demonstration to be a hoax. A trap-door beneath the chamber he was in led to a warm and comfortable room, Fox reported, while a body double took his place in the ice block. How did he get away with this trick? A lot of what magicians practice over and over again is getting the audience to accept things that are a bit out of the ordinary. There are some telltale clues that all was not as it seems. First, why is he wearing a mask? (You might assume that it's because it's part of the show, or because it makes him look fierce. The real reason might be because it makes it easier to fool you with a body double.) Why do they need to spray sheets of water over the ice at periodic intervals? (Blaine says it's to prevent the ice from melting; maybe it's so that he can change places with the body double during the brief moment you can't see through the ice.) What about the physiological monitoring equipment on his body, reporting his heart rate and body temperature—surely that's real, isn't it? (Who says that the equipment is actually hooked up to him? Perhaps it wasn't and was instead being fed by a computer.)

If Blaine was lying about the ice block—claiming it was a feat of endurance when in reality it was just conjuring, a magic trick—why not lie about other feats of endurance too? As a performer with a large audience, he would want to ensure that his demonstrations work every time. Using illusions and tricks may be more reliable, and safer, than trying to push endurance limits. But even if it did involve a trick, perhaps it's too harsh to call it a lie—it's all part of the show, isn't it? No one really believes that magicians are calling upon unseen forces; we know that they rehearse like the dickens and use misdirection. Who cares? Well, most reputable magicians, when asked, will come clean and admit that what they are doing are rehearsed illusions, not demonstrations of the black arts. Glenn Falkenstein, for example, performed a mind-reading act that was among the most impressive ever seen. But at the end of each show, he was quick to point out that there was no actual mind-reading involved. Why? Out of a sense of ethics. The world is full of people who believe things that aren't true, and believe many things that are ridiculous, he said. Millions of people who have a poor understanding of cause and effect waste their money and energy on psychics, astrologers, gambling, and "alternative" therapies with no proven efficacy. Being forthright about how this sort of entertainment is accomplished is important, he said, so that people are not led to believe things that aren't so.

In another demonstration, Blaine claims to have stuck a needle clear through his hand. Was this an illusion or did he really do it? In videos, it certainly looks real, but of course, that's what magic is all about. (Search YouTube and you'll find videos showing how it can be done with specialized apparatus.) What about the forty-four-day fast in a glass box? There was even a peer-reviewed paper in the *New England Journal of Medicine* about that, and in terms of information sources, that's about as good as it gets. Upon closer examination, however, the physicians who authored that paper only examined Blaine after the fast, not before or during, and so they can't provide independent verification that he actually fasted. Was this question ever raised during peer review? The current editor of the journal searched his office archives but the records had been destroyed, since the article was published a decade before my inquiry. The lead author on the article told me in an email that based on the hormones she measured after the event, he was indeed fasting, but it's possible as well that he was sneaking in some food; she couldn't comment on that. She did point me to an article by a colleague of hers in another peer-reviewed journal, in which a physician *did* monitor Blaine throughout the fast (the article didn't show up in my PubMed or Google Scholar searches because David Blaine was not mentioned in the article by name). Relevant is the following passage from the article, which appeared in the journal *Nutrition:*

Immediately before the start of the fast, DB appeared to have a muscular build that was consistent with the body mass index, body composition figures, and upper arm muscle circumference, which are reported below. On the evening of Saturday, September 6, 2003, DB entered a transparent Perspex box, measuring 2.1 x 2.1 x 0.9 m, which was suspended in air for the next 44 d, close to Tower Bridge, London. Continuous detailed video monitoring was available to one of the investigators (ME, office and at home), who was able to assess the clinical state and physical activity of DB. DB, who was 30 y old, had consumed before the event, a diet that was estimated, but not verified, to have increased his weight by as much as 6–7 kg. He also took some multivitamin tablets for a few days before the event, which he stopped on entry into the box. He felt weaker and more lethargic as the event progressed. From about 2 wk onward he experienced some dizziness and faintness on standing up quickly, and on some occasions, temporary visual problems, as if "blacking out." He also developed transient sharp shooting pains in his limbs and trunk, abdominal discomfort, nausea, and some irregular heart beats. . . . . A small amount of bleeding from his nose occurred on the fifth day after entry to the box and this recurred later. There were no other obvious signs or symptoms of a bleeding tendency. There were also no signs of edema before or at the end of the fast. In addition, there were no clinical signs of thiamine deficiency. DB, who was initially a muscular looking man, was visibly thinner on exit from the box. His blood pressure taken almost immediately before the event began was 140/90 mmHg while lying and 130/80 mmHg while standing, and at the end it was 109/74 mmHg while lying (pulse 89 beats/min) and 109/65 mmHg while standing (pulse 119 beats/min).

From this report, it does sound as though he really did fast. A skeptic might discount his reports of pain and nausea as showmanship, but it is difficult to fake irregular heartbeat and weight loss.

But it's the breath holding, televised on Oprah Winfrey's show, that is the focus of Blaine's TEDMED talk. In it, Blaine uses a lot of scientific and medical terminology to prop up the narrative that this was a medically based endurance demonstration, not a mere trick. Blaine describes the research he did:

"I met with a top neurosurgeon and I asked him how long . . . anything over six minutes you have a serious risk of hypoxic brain damage . . . perflubron." Blaine mentions liquid breathing; a hypoxic tent to build up red blood cell count; pure O_2. That got him to fifteen minutes. He goes on to elaborate a training regimen in which he gradually built up to seventeen minutes. He throws out terms like "blood shunting" and "ischemia." Did Blaine actually do what he said he did? Was the medical jargon he threw out for real, or just pseudoscientific babble he invoked to overwhelm, to make us *think* he knew what he was talking about?

As always, we start with a plausibility check. If you've ever tried holding your breath, you probably held out for half a minute—maybe even an entire minute. A bit of research reveals that professional pearl divers routinely hold their breaths for seven minutes. The world record for breath holding *before* Blaine's was just under seventeen minutes. As you continue to read up on the topic, you'll discover that there are two kinds of breath-holding competitions: plain old, regular old breath holding, like you and your older brother did in the community pool when you were kids, and *aided* breath holding, in which competitors are allowed to do things like inhale 100 percent pure oxygen for half an hour prior to competing. This is sounding more plausible, but how far can you get with aided breath holding—can it actually bridge the gap between a few minutes and seventeen minutes? At this point, you might try to learn what the experts have to say—pulmonologists (who would know something about lung capacity and the breathing reflex) and neurologists (who would know how long the brain can last without an influx of oxygen). The two pulmonologists I checked with described a training regimen much like the one Blaine describes in his video; both felt that with these "tricks" or special measures, seventeen minutes of breath holding would be possible. Indeed, Blaine's record was broken in 2012 by Stig Severinsen, who held his breath for twenty minutes and ten seconds (after inhaling pure oxygen, of course), and who then broke his own record a month later, achieving twenty-two minutes. David Eidelman, MD, a pulmonary specialist and dean of the McGill Medical School said: "I agree that it does sound hard to believe. . . . However, by inhaling oxygen first, fasting, and using yoga-type techniques to lower metabolic rate while holding one's breath underwater, it seems that this is possible. So, while I retain some skepticism, I do not think I can prove it is impossible."

Charles Fuller, MD, a pulmonary specialist at UC Davis, adds, "There is sufficient evidence to indicate that Blaine is being truthful, as this event is physiologically feasible. Given the caveat that Blaine is a magician and there could have been other contributing factors to his successful seventeen-minute breath hold, there is also ample physiological evidence that this feat could have been accomplished. There is a subset of people in the breath-hold world who vie for a record officially known as 'pre-oxygenated static apnea.' In this event, breath holding is sponsored by Guinness World Records, as sports divers consider this cheating. Breath-hold duration is measured after hyper-ventilating (blowing of carbon dioxide) for thirty minutes while breathing 100 percent pure oxygen. Further, the event is typically held in a warm pool (which reduces metabolic oxygen demand), with the head held just below the surface, which induces the human dive reflex (further depressing metabolic oxygen demand). In other words, all 'tricks' which extend the human capacity for conscious breath hold. Most importantly, prior to Blaine, the record was just under his sev-enteen minutes [by an athlete who was *not* a magician], and there are additional individuals who have now been recorded for longer breath holds in excess of twenty minutes. Thus, ample evidence that this feat could have been accomplished as claimed."

So far, Blaine's story seems plausible, and his talk hits all the right notes. But what about brain damage? Blaine himself men-tioned this as a problem. You've no doubt heard that if the brain loses oxygen for even three minutes, irreparable damage and brain death can occur. If you're not breathing for seventeen minutes, how do you prevent brain death? A good question for a neurologist.

Scott Grafton, MD: "Oxygen doesn't stay in the blood all by itself. Think oil and water. It will quickly diffuse out of the liquid blood—it needs to bind to something. Blood carries red cells. Each red cell is loaded with hemoglobin (Hgb) molecules. These hemoglobin mol-ecules can potentially bind up to four oxygen molecules. Each time a red cell passes through the lungs, the number of Hgb molecules with oxygen bound to them increases. The stronger the concentration of oxygen in the air, the more Hgb molecules will bind to it. So load 'em up! Breathe 100 percent oxygen for thirty minutes so that total oxygen binding is as close to 100 percent saturation as you can get.

"Each time the red cell passes through the brain, oxygen will have a probability of unbinding from the molecule, diffusing across cell membranes to enter the brain tissue, where it binds to other molecules that use it in oxidative metabolism. The probability of a given oxygen molecule unbinding from hemoglobin and diffusing is a function of the relative difference of oxygen concentration on either side of the membranes."

In other words, the more oxygen the brain needs, the more likely it will be to pull oxygen out of hemoglobin. By breathing pure oxygen for thirty minutes, the competitive breath holder will maximize the amount of oxygen in the brain *and* in the blood. Then, once the breath-holding event starts, oxygen levels in the brain will decrease as they normally do over time, and the competitive breath holder will very efficiently pull out whatever oxygen happens to be left in hemoglobin to oxygenate the brain.

Grafton continues, "Not all the hemoglobin molecules are loaded up with oxygen on each pass through the lungs, and not all of them unload on each pass through the organs. It takes quite a few passes to unload all of them. When we say that brain death happens quickly due to a lack of oxygen, it is usually in the context of a lack of circulation (heart attack) when the heart is no longer delivering blood to the brain. Stop the pump, and no red cells are available to offer up oxygen and brain tissue dies fast. In a person submerged, there is a race between brain injury and pump failure.

"One key trick: The muscles need to be at rest. Muscles are loaded with myoglobin, which holds on to oxygen four times more strongly than red-cell hemoglobin does. If you're using your muscles, this will accelerate the loss of oxygen overall. Keep muscle demand low." This is the *static* in the static apnea that Dr. Fuller mentioned.

So from a medical standpoint, David Blaine's claim appears plausible. That might be the end of the story, except for this. An article in the *Dallas Observer* claims the breath holding was a trick, and that Blaine—a master illusionist—used a well-hidden breathing tube. There's nothing about this in other mainstream media, which doesn't mean the *Observer* is wrong, of course, but why is this paper the only one reporting it? Perhaps a magician who performs a trick but claims it wasn't one is not big news.

Reporter John Tierney traveled to Grand Cayman Island to write about Blaine's preparation for the breath holding for an article in the *New York Times,* and then wrote about the *Oprah* appearance in his blog a week later. Tierney makes a lot out of Blaine's heart rate, as reported on a monitor next to his tank on the Winfrey show, but as with the ice-block demonstration, there's no evidence this monitor was actually connected to Blaine, and it might really have been more for showmanship—to make the audience think that the conditions were really rough (a standard practice for magicians). Neither Tierney nor a physician involved in the training mention how closely they were monitoring Blaine during practice trials in the Caymans—it's possible that they took him at his word that he didn't have any apparatus. Perhaps the real motive of this training was that Blaine figured if he could fool *them,* he could fool a television audience. Tierney writes, "I was there at the pool along with some free-divers who are experts at static apnea (holding your breath while remaining immobile). Dr. Ralph Potkin, a pulmonologist who studies breath holding and is the team physician for the United States free-diving team, attached electrodes to Blaine's body during the session and measured his heart, blood, and breathing as Blaine kept his head submerged in the water for sixteen minutes.

"I've always been skeptical of cons—I did a long piece on James Randi a while back, and was with him in Detroit when he was exposing an evangelist named Peter Popoff—but I saw no reason to doubt Blaine's feat. His breath hold in front of me was done in clear water in the shallow end of the very ordinary swimming pool at our hotel, with experts in breath holding a few feet away watching him all the time. His nose and mouth were clearly below the water—but just a couple of inches, so they were visible at all times. You tell me how he snuck a breathing tube in there so that no one noticed it or any bubbles. Magicians fool people by distracting them with motions and patter, but the whole point of static apnea is to remain absolutely motionless in order to conserve oxygen, which is what David did. (It's remarkable what a difference that makes—the trainers who were working with David did a short session with me and my photographer. We pre-breathed air instead of oxygen, but we were amazed at how long we went—I got up to 3 minutes and 41 seconds, and the photographer even longer.)"

So now the *Dallas Observer* says it was faked, and a *New York Times* reporter seems to believe it wasn't. What do professional magicians think? I spoke to four. One said, "It's got to be a trick. A lot of his demonstrations are known, at least within the magic community, to use camera trickery and [a] very involved setup. It would be very easy for him to have a breathing tube that allows him to take oxygen in, and to exhale carbon dioxide, without making bubbles in the water. And if he practices, he wouldn't need to be doing it all that often—he could actually hold his breath for a minute or two at a time in between tube breaths. And there could be other camera trickery—he might not actually be in the water! Projection or green screen could make it appear that he was."

The second magician, who had worked with Blaine a decade earlier, added, "His hero is Houdini, who became famous for doing stunts. Houdini made a reputation in part by doing things that people did in the 1920s—flagpole sitting and so on. Some do require endurance and some are faked slightly; some aren't as hard as you'd think, but most people never try them. I don't see why Blaine would fake the block-of-ice trick—that one is simple because of the igloo effect—it's not actually that cold in there. It *looks* impressive. If he was in a freezer that would be different.

"But seventeen-minute breath holding? If he can super-oxygenate his blood, that can help. I know that he does train and does some things that are remarkable. But I'm sure the breath-holding trick is partly enabled. He does hold his breath, I think, but not 100 percent of the time. It's quite easy to fake. He probably has [a] breathing tube and other apparatus.

"Note that a lot of his magic is on TV and there are edits at key points. We assume it's real information and we're seeing everything because that's how our brains construct reality. But as a magician, I see the edits and wonder what was happening during the missing footage."

A third magician added, "Why would you go to all the trouble to train if, as an illusionist, you can do it with equipment? Using equipment creates a more reliable, replicable performance that you can do again and again. Then, you just need to act as though you're in pain, dizzy, disoriented, and as though you've pushed your body beyond all reasonable limits. As an entertainer, you wouldn't want to leave anything to chance—there's too much at stake."

The fact that no one reports seeing David Blaine use a breathing tube does not constitute evidence that he didn't, because it is precisely the *job* of illusionists to play tricks with what you see and what you think you see. And the illusion is even more powerful when it happens to you. I've had the magician Glenn Falkenstein read off the serial numbers of dollar bills in my wallet from across the room while he was blindfolded. I've had the magician Tom Nixon place the seven of diamonds in my hand, yet a few minutes later it had become a completely different card without my being aware of him touching me or the card. I know he's switching the card at some point, but even after having the trick done to me five times, and watching it performed on other people many more times, I still don't know when the switch occurs. That is part of the magician's genius, and part of the entertainment. I don't think for a moment that Falkenstein or Nixon possess occult powers. I know it's entertainment, and they sell it as that.

The fourth magician I asked about it was James Randi, the professional skeptic I (and John Tierney) mentioned earlier, who replicates alleged psychic phenomena through his deft use of illusions and magic tricks. Here's what he wrote via email:

> I recall that when David Blaine first showed up on television performing his stunts, I voluntarily contacted him with a friendly warning that he was—in my opinion as a conjuror—taking chances of personal physical damage. We exchanged friendly correspondence on this matter, until I was abruptly informed that his newly engaged management agency had changed his email address and that he'd been instructed not to correspond further with me. I of course accepted this decision, while hoping that Mr. Blaine would heed my well-intended suggestions.
>
> I have not been in touch with David Blaine since that time. I was alarmed to see the unwise statements he made on the TED appearance, and I have respected the—to me—rather unwise slant that his agency has chosen to give to his claims, but I have respected his privacy.
>
> He let his agency even terminate his connection with me, perhaps because I might have tried to keep him honest. Can't have too much of that quality, of course.

The weight of our fact-checking suggests that the seventeen-minute breath hold is very plausible. That doesn't guarantee that

Blaine didn't use a breathing tube. Whether you believe Blaine pulled off the stunt legitimately is up to you—each of us has to make our own decision. As with any magician, we can't be sure what's true and what's not—and that is the world of ambiguity that magicians spend their professional lives trying to create. In critical thinking, one looks for the most parsimonious account, but in some cases, as here, it is difficult or impossible to choose between the possible explanations or to figure which is more parsimonious. Does it even matter? Well, yes. As Falkenstein said, people who have a poor understanding of cause and effect, or an insufficient understanding of chance and randomness, are easily duped by claims such as these, leading them to too readily accept others. Not to mention the many amateurs who may try to replicate these spectacles, despite the ubiquitous warning of "do not try this at home." The uneducated are easy targets. The difference between doing this by training and doing it by illusion is the difference between being duped and not being duped.

Statistics in the Universe

When you hear names like hydrogen, oxygen, boron, tin, and gold, what do you think of? These are chemical elements of the periodic table, usually taught in middle or high school. They were called elements by scientists because they were believed to be fundamental, indivisible units of matter (from the Latin *elementum*, matter in its most basic form). The Russian scientist Dmitri Mendeleev noticed a pattern in the properties of elements and organized them into a table that made these properties easier to visualize. In the process, he was able to see gaps in the table for elements that had not yet been discovered. Eventually all of the elements between 1 and 118 have either been discovered in nature or synthesized in the laboratory, supporting the theory underlying the table's arrangement.

Later, scientists discovered that the chemical elements were not actually indivisible; they were made of something that the scientists called atoms, from the Greek word *atomos*, for "indivisible." But they were wrong about the indivisibility of those, too—atoms were later discovered to be made up of subatomic particles: protons, neutrons, and electrons. These were also initially thought to be indivisible, but then—you guessed it—that was found to be incorrect. The so-called Standard Model of Particle Physics was formulated in the 1950s and '60s, and theorized that electrons are indivisible, but protons and neutrons are composed of smaller subatomic particles. With the discovery of quarks in the 1970s, this model was confirmed. To further complicate terminology, protons and electrons are a type of *fermion*, and neutrons are a type of *boson* (photons are also a type of boson). The different categories are necessary because the two different types of particles are governed by different laws. Fermions and bosons have been given the name *elementary particle* because it is believed that they are truly indivisible (but time will tell).

According to the Standard Model, there are seventeen different types of elementary particles—twelve kinds of fermions and five kinds of bosons. The Higgs boson, which received a great deal of press in 2012 and 2013, has been the last remaining piece of the Standard Model to be proven—the other sixteen have already been discovered. If it exists, the Higgs would help to explain how matter obtains mass, and fill in a key hole in the theory used to explain the nature of the universe, a hole in the theory that has existed for more than fifty years.

How do we know if we've found it? When particles collide at great . . . Oh, forget it. I'll let a physicist explain it. Here's Professor Harrison Prosper, describing this plot and the little "blip" next to the arrow corresponding to 125 gigaelectronvolts (GeV) on the horizontal axis:

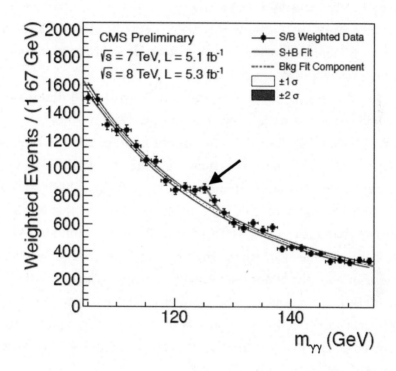

The graph shows "a spectrum arising from proton-proton collisions that resulted in the creation of a pair of photons (gammas in high energy argot)," Prosper says. "The Standard Model predicts that the Higgs boson should decay (that is, break up) into a pair of photons. (The Higgs is predicted to decay in other ways too, such as a pair of Z bosons.) The bump in the plot at around 125 GeV is evidence for the existence of some particle of a definite mass that decays into a pair of photons. That something, as far as we've been able to ascertain, is likely to be the Higgs boson."

Not all physicists agree that the experiments are conclusive. Louis Lyons explains, "The Higgs . . . can decay to different sets of particles, and these rates are defined by the S.M. [Standard Model]. We measure these ratios, but with large uncertainties with the present data. They are consistent with the S.M. predictions, but it could be much more convincing with more data. Hence the caution about saying we have discovered the Higgs of the S. M."

In other words, the experiments are so costly and difficult to conduct, that physicists want to avoid a false alarm—they've been wrong before. Although CERN officials announced in 2012 that they had found it, many physicists feel the sample size was too small. There is so much at stake that the physicists have set for themselves a standard of proof, a statistical threshold, that is much stricter than the 1 in 20 used in other fields—1 in 3.5 million. Why such an extreme evidence requirement? Prosper says, "Given that the search for the Higgs took some forty-five years, tens of thousands of scientists and engineers, billions of dollars, not to mention numerous divorces, huge amounts of sleep deprivation, tens of thousands of bad airline meals, etc., etc., we want to be sure as is humanly possible that this is real."

Physicist Mads Toudal Frandsen adds, "The CERN data is generally taken as evidence that the particle is the Higgs particle. It is true that the Higgs particle can explain the data but there can be other explanations; we would also get this data from other particles. The current data is not precise enough to determine exactly what the particle is. It could be a number of other known particles." Recall the discussion earlier in the *Field Guide* about alternative explanations. Physicists are on the alert for this.

If the plot is showing evidence of a different kind of particle, something that is *not* the Higgs, this could substantially change our view of how the universe was created. And if it does exist, some physicists, such as Stephen Hawking, fear that this could spell the end of the universe as we know it. The fear is that a quantum fluctuation could create a vacuum bubble that rapidly and continually expands until it wipes out the universe. And if you think physicists don't have a sense of humor, Joseph Lykken, a physicist and director of the Fermi National Accelerator Laboratory in Illinois, noted that it won't happen for a long, long time—10^{100} years from now—"so probably you shouldn't sell your house and you should continue to pay your taxes."

Not everyone is happy with the discovery, and not because it may signal the end of the world—it's because finding something in science that the standard theories predict doesn't open the door for new inquiry. An anomalous, unexplained result is most interesting to scientists because it means their model and understanding was at best incomplete, and at worst, completely wrong—presenting a great opportunity for new learning. In one of the many intersections between art and science, the conductor Benjamin Zander says that when a musician makes a mistake, rather than swearing or saying "oops" or "I'm sorry," she should say, "Now *that's* interesting!" Interesting because it represents an opportunity for learning. It could be that the discovery of the Higgs boson answers all the questions we had. Or, as *Wired* writer Signe Brewster says, "It could lead to an underlying principle that physicists have missed until now. The end goal, as always, is to find a string that, when tugged, rings a clarion bell that draws physicists toward something new." As Einstein reportedly said, if you know how it's going to turn out, it's not science, it's engineering.

Scientists are curious, lifelong learners, eager to find the next challenge. There are some who fear that the discovery of the Higgs may explain so much that it ends the ride. Others are so filled with wonder and the complexities of life and the universe that they are confident we'll never figure it all out. I am among the latter.

As of this writing, tantalizing evidence has emerged from CERN of a new particle that might be a graviton, or a heavier version of the Higgs boson. But the most probable explanation for these surprising new bumps in the data flow is that it is a coincidence—the findings have a 1 in 93 chance of being a fluke, far more likely than the 1 in 3.5 million probability used for the Higgs. But there are qualitative considerations. "What is nice is that it is not a particularly crazy signal, in a quite clean channel," physicist Nima Arkani-Hamed told the *New York Times*. "So, while we are nowhere near moving champagne even vaguely close to the fridge, it is intriguing." Nobody knows yet what it is, and that's just fine with Lykken and many others who love the thrill of the chase.

Science, history, and the news are full of things that we knew, or thought we did, until we discovered we were wrong. An essential component of critical thinking is knowing what we don't know. A guiding principle, then, is simply that we know what we know until we don't any longer. The purpose of the *Field Guide* was to help you to think things through, and to give you greater confidence both in what you think you know, and what you think you don't, and—hopefully—to be able to tell the difference between them.

CONCLUSION

DISCOVERING YOUR OWN

In George Orwell's *1984*, the Ministry of Truth was the country's official propaganda agency, charged with falsifying historical records and other documents to reflect the administration's agenda. The Ministry also advanced counterknowledge when it served their purposes, such as $2 + 2 = 5$.

Nineteen Eighty-four was published in 1949, half a century before the Internet became our de facto information source. Today, like in *1984*, websites can be altered so that the average person doesn't know that they have been; every trace of an old piece of information can be rewritten, or (in the case of Paul McCartney and Dick Clark) kept out of reach. Today, it can be very difficult for the average Web surfer to know if a site is reporting genuine knowledge or counterknowledge. Unfortunately, sites that advertise that they are telling the truth are often the ones that aren't. In many cases, the word "truth" has been co-opted by people who are propagating counterknowledge or fringe viewpoints that go against what is conventionally accepted as truth. Even site names can be deceptive.

Can we trust experts? It depends. Expertise tends to be narrow. An economist in the highest echelons of the government may not have any special insight into what social programs will be effective for curbing crime. And experts sometimes become co-opted by special interests, and, of course, they make mistakes.

An anti-science bias has entered public discourse and the Web. A lot of things that should be scientific or technical problems—like where to put a power plant and how much it should cost—are political. When that happens, the decision-making process is subverted, and the facts that matter are often not the ones that are under consideration. Or we say that we want to cure an intractable human disease, but mock the first step when tens of millions of dollars are spent studying aphids. The reality is that science progresses by gaining an understanding of basic cellular physiology. With the wrong frame the research looks trivial; with the right frame it can be seen for the potential it truly has to be transformative. Money put into human clinical trials might end up being able to treat the symptoms of a few hundred thousand people. That same money put into basic-level scientific research has the potential to find the cure for *dozens* of diseases and *millions* of people because it is dealing with mechanisms common to many different types of bacteria and viruses. The scientific method is the ground from which all the best critical thinking rises.

In addition to an anti-science bias, there is an anti-skepticism bias when it comes to the Internet. Many people think, "If I found it online it must be true." With no central authority charged with the responsibility of monitoring and regulating websites and other material found online, the responsibility for verifying claims falls on each of us. Fortunately, some websites have cropped up that help. Snopes.com and similar sites are dedicated to exposing urban legends and false claims. Companies such as Consumer Reports run independent laboratories to provide an unbiased assessment of different products, regardless of what their manufacturers claim. Consumer Reports has been around for decades, but it is no great leap to expect that other critical-thinking enterprises will flourish in the twenty-first century. Let's hope so. But whatever helpful media is out there, each of us will still have to apply our judgment.

The promise of the Internet is that it is a great democratizing force, allowing everyone to express their opinions, and everyone to have immediate access to all the world's information. Combine these two, as the Internet and social media do, and you have a virtual world of information and misinformation cohabiting side by side, staring back at you like identical twins, one who will help you and the other who will hurt you. Figuring out which one to choose falls upon all of us, and it requires careful thinking and one thing that most of us feel is in short supply: time. Critical thinking is not something you do once with an issue and then drop it. It's an active and ongoing process. It requires that we all think like Bayesians, updating our knowledge as new information comes in.

Time spent evaluating claims is not just time well spent, it should be considered part of an implicit bargain we've all made. Information gathering and research that used to take anywhere from hours to weeks now takes just seconds. We've saved incalculable numbers of hours of trips to libraries and far-flung archives, of hunting through thick books for the one passage that will answer our questions. The implicit bargain that we all need to make explicit is that we will use just *some* of that time we saved in information acquisition to perform proper information verification. Just as it's difficult to trust someone who has lied to you, it's difficult to trust your own knowledge if half of it turns out to be counterknowledge. The fact is that right now counterknowledge flourishes on Facebook and on Twitter and on blogs . . . on all the semi-organized platforms.

We're far better off knowing a moderate number of things with certainty than a large number of things that might not be so. Counterknowledge and misinformation can be costly, in terms of lives and happiness, and in terms of the time spent trying to undo things that didn't go the way we thought they would. True knowledge simplifies our lives, helping us to make choices that increase our happiness and save time. Following the steps in this *Field Guide* to evaluate the myriad claims we encounter is how we can stay two steps ahead of the millions of lies that are out there on the Web, and ahead of the liars and just plain incompetents who perpetrate them.

APPENDIX
APPLICATION OF BAYES'S RULE

Bayes's rule can be expressed as follows; $P(A \mid B) = \dfrac{(P(B \mid A) \times P(A))}{(P(B))}$

For the current problem, let's use the notation that G refers to the prior probability that the suspect is guilty (before we know anything about the lab report) and E refers to the evidence of a blood match. We want to know P(G|E). Substituting in the above, we put in G for A and E for B to obtain:

$$P(G \mid E) = \frac{(P(E \mid G) \times P(G))}{(P(E))}$$

To compute Bayes's rule and solve for $P(G \mid E)$, it may be helpful to use a table. The values here are the same as those used in the fourfold table on page 220.

COMPUTATION OF BAYES'S RULE

Hypothesis (H) (1)	Prior Probability P(G) (2)	Evidence Probability P(E | G) (3)	Product (4) = (2)(3)	Posterior Probabilities P(G | E) (6) = (4)/Sum
Guilty	.02	.85	.017	.104
Innocent	.98	.15	.147	.896

Sum = .164
= P(D)

Then, rounding, P(Guilty | Evidence) = .10
P(Innocent | Evidence) = .90

GLOSSARY

This list of definitions is not exhaustive but rather a personal selection driven by my experience in writing this book. Of course, you may wish to apply your own independent thinking here and find some of the definitions deserve to be challenged.

Abduction. A form of reasoning, made popular by Sherlock Holmes, in which clever guesses are used to generate a theory to account for the facts observed.

Accuracy. How close a number is to the true quantity being measured. Not to be confused with precision.

Affirming the antecedent. Same as *modus ponens* (see entry below).

Amalgamating. Combining observations or scores from two or more groups into a single group. If the groups are similar along an important dimension—homogeneous—this is usually the right thing to do. If they are not, it can lead to distortions of the data.

Average. This is a summary statistic, meant to characterize a set of observations. "Average" is a nontechnical term, and usually refers to the *mean* but could also refer to the *median* or *mode.*

Bimodal distribution. A set of observations in which two values occur more often than the others. A graph of their frequency versus their values shows two peaks, or humps, in the distribution.

Conditional probability. The probability of an event occurring *given* that another event occurs or has occurred. For example, the probability that it will rain today *given* that it rained yesterday. The word "given" is represented by a vertical line like this: | .

Contrapositive. A valid type of deduction of the form:

If A, then B
Not B
Therefore, not A

Converse error. An invalid form of deductive reasoning of the form:

> If A, then B
> B
> Therefore, A

Correlation. A statistical measure of the degree to which two variables are related to each other, it can take any value from −1 to 1. A perfect correlation exists (correlation = 1) when one variable changes perfectly with another. A perfect negative correlation exists when one variable changes perfectly opposite the other (correlation = −1). A correlation of 0 exists when two variables are completely unrelated.

A correlation shows only that two (or more) variables are linked, not that one causes the other. Correlation does not imply causation.

A correlation is also useful in that it provides an estimate for how much of the variability in the observations is caused by the two variables being tracked. For example, a correlation of .78 between height and weight indicates that 78 percent of the differences in weight across individuals are linked to differences in height. The statistic doesn't tell us what the remaining 22 percent of the variability is attributed to—additional experimentation would need to be conducted, but one could imagine other factors such as diet, genetics, exercise, and so on are part of that 22 percent.

Cum hoc, ergo propter hoc (**with this, therefore because of this**). A logical fallacy that arises from thinking that just because two things co-occur, one must have caused the other. Correlation does not imply causation.

Cumulative graph. A graph in which the quantity being measured, say sales or membership in a political party, is represented by the total to date rather than the number of new observations in a time period. This was illustrated using the cumulative sales for the iPhone [page 48].

Deduction. A form of reasoning in which one works from general information to a specific prediction.

Double y-axis. A graphing technique for plotting two sets of observations on the same graph, in which the values for each set are represented on two different axes (typically with different scales). This is only appropriate when the two sets of observations are measuring unlike quantities, as in the graph on page 40. Double y-axis graphs can be misleading because the graph maker can adjust the scaling of the axes in order to make a particular point. The example used in the text was a deceptive graph made depicting practices at Planned Parenthood.

Ecological fallacy. An error in reasoning that occurs when one makes inferences about an individual based on aggregate data (such as a group mean).

Exception fallacy. An error in reasoning that occurs when one makes inferences about a group based on knowledge of a few exceptional individuals.

Extrapolation. The process of making a guess or inference about what value(s) might lie beyond a set of observed values.

Fallacy of affirming the consequent. See *Converse error.*

Framing. The way in which a statistic is reported—for example, the context provided or the comparison group or amalgamating used—can influence one's interpretation of a statistic. Looking at the total number of airline accidents in 2016 versus 1936 may be misleading because there were so many more flights in 2016 versus 1936—various adjusted measures, such as accidents per 100,000 flights or accidents per 100,000 miles flown, provide a more accurate summary. One works to find the true frame for a statistic, that is the appropriate and most informative one. Calculating proportions rather than actual numbers often helps to provide the true frame.

GIGO. Garbage in, garbage out.

Incidence. The number of new cases (e.g., of a disease) reported in a specified period of time.

Inductive. A form of inferential reasoning in which a set of particular observations leads to a general statement.

Interpolation. The process of estimating what intermediate value lies between two observed values.

Inverse error. An invalid type of deductive reasoning of the form:

> If A, then B
> Not A
> Therefore, not B

Mean. One of three measures of the average (the central tendency of a set of observations). It is calculated by taking the sum of all observations divided by the number of observations. It's what people usually are intending when they simply say "average." The other two kinds of averages are the median and the mode. For example, for {1, 1, 2, 4, 5, 5} the mean is (1 + 1 + 2 + 4 + 5 + 5) ÷ 6 = 3. Note that, unlike the mode, the mean isn't necessarily a value in the original distribution.

Median. One of three measures of the average (the central tendency of a set of observations). It is the value for which half the observations are larger and half are smaller. When there is an even number of observations, statisticians may take the mean of the two middle observations. For example, for

{10, 12, 16, 17, 20, 28, 32}, the median is 17. For {10, 12, 16, 20, 28, 32}, the median would be 18 (the *mean* of the two middle values, 16 and 20).

Mode. One of three measures of the average (the central tendency of a set of observations). It is the value that occurs most often in a distribution. For example, for {100, 112, 112, 112, 119, 131, 142, 156, 199} the mode is 112. Some distributions are bimodal or multimodal, meaning that two or more values occur an equal number of times.

Modus ponens. A valid type of deductive argument of the form:

If A, then B
A
Therefore, B

Post hoc, ergo propter hoc **(after this, therefore because of this).** A logical fallacy that arises from thinking that just because one thing (Y) occurs after another (X), that X *caused* Y. X and Y might be correlated, but that does not mean a causative relation exists.

Precision. A measure of the level of resolution of a number. The number 909 is precise to zero decimal points and has only a resolution of the nearest whole number. The number 909.35 is precise to two decimal places and has a resolution of 1/100th of a unit. Precision is not the same as accuracy—the second number is more precise, but if the true value is 909.00, the first number is more accurate.

Prevalence. The number of existing cases (e.g., of a disease).

Scatter plot. A type of graph that represents all points individually. For example, opposite is a scatter plot of the data presented on page 39.

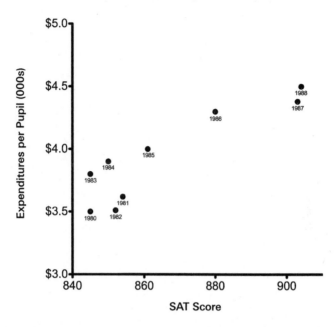

School Funding v. SAT Scores

Subdividing. Breaking up a set of observations into smaller groups. This is acceptable when there is heterogeneity within the data and the larger group is composed of entities that vary along an important dimension. But subdividing can be used deceptively to create a large number of small groups that do not differ appreciably along a variable of interest.

Syllogism. A type of logical statement in which the conclusion must necessarily follow from the premises.

Truncated axis. Starting an x- or y-axis on a value other than the lowest one possible. This can sometimes be helpful in allowing viewers to see more clearly the region of the graph in which the observations occur. But used manipulatively, it can distort the reality. The graph shown in this glossary under the entry for "scatter plot" uses two truncated axes effectively and does not give a false impression of the data. The graph shown on page 29 by Fox News does give a false impression of the data, as shown in the redrawn version on page 30.

NOTES

INTRODUCTION: THINKING, CRITICALLY

x **avoid learning a whole lot of things that aren't so** After Huff, D. (1954/1993). *How to Lie with Statistics*. New York: W.W. Norton, p. 19. And, as you'll read later, he probably was echoing Mark Twain, or Josh Billings, or Will Rogers, or who knows who.

xi **Misinformation has been a fixture of human life** Abraham provides misinformation about the identity of his wife, Sarah, to King Abimelech to protect himself. The Trojan horse was a kind of misinformation, appearing as a gift but containing soldiers.

PART ONE: EVALUATING NUMBERS

3 **People choose what to count** This sentence is nearly a direct quote from Best, J. (2005). Lies, calculations and constructions: beyond *How to Lie with Statistics*. *Statistical Science, 20*(3), 210–214.

5 **More people have cell phones than toilets** Wang, Y. (2013, March 25). More people have cell phones than toilets, U.N. study shows. http://news feed.time.com/2013/03/25/more-people-have-cell-phones-than-toilets-u-n -study-shows/.

6 **150,000 girls and young women die of anorexia each year** Steinem, G. (1992). *Revolution from Within*. New York: Little, Brown. Wolf, N. (1991). *The Beauty Myth*. New York: William Morrow.

6 **Add in women from twenty-five to forty-four and you still only get 55,000** This example came to my attention from Best, J. (2005). Lies, calculations and constructions: beyond *How to Lie with Statistics*. *Statistical Science, 20*(3), 210–214. The statistics are available at www.cdc.gov.

6 **anorexia deaths in one year cannot be three times the number of *all* deaths** Maybe you're in the accounts payable department of a big

corporation. An employee put in for reimbursement of gasoline for the business use of his car, $5,000 for the month of April. Start with a little world knowledge: Most cars get better than twenty miles per gallon these days (some get several times that). You also know that the fastest you can reasonably drive is seventy miles per hour, and that if you were to drive ten hours a day, all on the freeway, that would mean 700 miles a day. Keep that up for a standard 21.5-day work month and you've got 15,050 miles. In these kinds of rough estimates, it's standard to use round numbers to make things easier, so let's call that 15,000. Divide that by the fuel economy of 20 mpg and, by a rough estimate, your employee needed 750 gallons of gas. You look up the average national gas price for April and find that it's $2.89. Let's just call that $3.00 (again, rounding, and giving your employee the benefit of the doubt—he may not have managed to get the very best price every time he filled up). $3/gallon times 750 gallons = $2,250. The $5,000 on the expense report doesn't look even remotely plausible now. Even if your employee drove twenty hours a day, the cost wouldn't be that high. https://www.fueleconomy.gov/feg/best/bestworstNF.shtml, retrieved August 1, 2015. http://www.fuelgaugereport.com/.

6 **a telephone call has decreased by 12,000 percent** Pollack, L., & Weiss, H. (1984). Communication satellites: countdown for Intelsat VI. *Science, 223*(4636), 553.

6 **one of 12 percent seems wildly unlikely** I suppose you could spin a story that makes this true. Maybe a widget used to cost $1, and now, as part of a big promotion, a company is not just willing to give it to you for free, but to *pay* you $11,999 to take it (that's a 12,000 percent reduction). This happens in real estate and big business. Maybe an old run-down house needs to be razed before a new one can be built; the owner may be paying huge property taxes, the cost of tearing down the house is high, and so the owner is willing to pay someone to take it off of his or her hands. At one point in the late 1990s, several large, debt-ridden record companies were "selling" for $0, provided the new owner would assume their debt.

6 **200 percent reduction in customer complaints** Bailey, C., & Clarke, M. (2008). Aligning business leadership development with business needs: the value of discrimination. *Journal of Management Development, 27*(9), 912–934.

Other examples of a 200 percent reduction: Rajashekar, B. S., & Kalappa, V. P. (2006). Effects of planting seasons on seed yield & quality of tomato varieties resistant to leaf curl virus. *Seed Research, 34*(2), 223–225. http://www.bostoncio.com/AboutRichardCohen.asp.

7 **50 percent reduction in salary** Illustration © 2016 by Dan Piraro based on an example from Huff, ibid.

7 **making this distinction between percentage point and percentages clear**
I'm grateful to James P. Scanlan, attorney-at-law, Washington, D.C., who
answered my query to the membership of the American Statistical Association, and provided me with this misuse.

8 **closing of a Connecticut textile mill and its move to Virginia** This example comes from Spirer, L., Spirer, H. F., & Jaffe, A. J. (1987). *Misused Statistics*, New York: Marcel Dekker, p. 194.

Miller, J. (1996, Dec. 29). High costs are blamed for the loss of a mill.
New York Times, Connecticut Section.

And n. a. (1997, Jan. 12). Correction, *New York Times*, Connecticut Section.

8 **legislation that denied additional benefits** McLarin, K. J. (1993, Dec. 5).
New Jersey welfare's give and take; mothers get college aid, but no extra
cash for newborns. *New York Times.*

See also: Henneberger, M. (1995, April 11). Rethinking welfare: deter-
ring new births—a special report; state aid is capped, but to what effect?
New York Times.

8–9 **births to welfare mothers had already fallen by 16 percent** Ibid.

9 **no reason to report the new births** Ibid.

16 **Although they are mathematically equivalent** Koehler, J. J. (2001). The
psychology of numbers in the courtroom: how to make DNA-match statistics seem impressive or insufficient. *Southern California Law Review, 74,*
1275–1305.

And Koehler, J. J. (2001). When are people persuaded by DNA match
statistics? *Law and Human Behavior, 25*(5), 493–513.

17 **On average, humans have one testicle** Attributed to mathematics professor Desmond MacHale of University College, Cork, Ireland.

17 **temperatures ranging from 15 degrees to 134 degrees** http://en.wikipedia
.org/wiki/Death_Valley.

18 **the amount of money spent on lunches in a week** As an example, suppose
six adults spend the following amounts on lunch {$12, $10, $10, $12, $11, $11}
and six children spend the following {$4, $3.85, $4.15, $3.50, $4.50, $4}. The
median (for an even number of observations, the median is sometimes
taken as the mean between the two middle numbers, or in this case, the
mean of 4.5 and 10) is $7.25. The mean and median are amounts that no
one actually spends.

19 **During the 2004 U.S. presidential election** See Gelman, A. (2008). *Red
State, Blue State, Rich State, Poor State.* Princeton, NJ: Princeton University Press.

20 **the average life expectancy for males and females** These numbers are for
white males and females. Non-white figures for 1850 are not as readily

available. http://www.infoplease.com/ipa/A0005140.html. An additional source of concern is that the U.S. numbers for 1850 are for the state of Massachusetts only, according to the Bureau of the Census.

21 **the average family** The title of this section, and the discussion, follows the work of Jenkins and Tuten very closely:

Jenkins, J., & Tuten, J. (1992). Why isn't the average child from the average family? And similar puzzles. *American Journal of Psychology, 105*(4), 517–526.

22 **the average number of siblings** Stick-figure children from Etsy, https://www.etsy.com/listing/221530596/stick-figure-family-car-van-bike-funny; small and large house drawn by the author; medium house from http://www.clipartbest.com/clipart-9TRgq8pac.

24 **average investor does not earn the average return** A simulation, see Tabarrok, A. (2014, July 11). Average stock market returns aren't average. http://marginalrevolution.com/marginalrevolution/2014/07/average-stock-market-returns-arent-average.html. Accessed October 14, 2014.

26 **poster presented at a conference by a student researcher** Tully, L. M., Lincoln, S. H., Wright, T., & Hooker, C. I. (2013). Neural mechanisms supporting the cognitive control of emotional information in schizophrenia. Poster presented at the 25th Annual Meeting of the Society for Research in Psychopathology. https://www.researchgate.net/publication/266159520 _Neural_mechanisms_supporting_the_cognitive_control_of_emotional _information_in_schizophrenia.

I first found this example at www.betterposters.blogspot.com.

27 **gross sales of a publishing company** http://pelgranepress.com/index.php /tag/biz/.

29 **Fox News broadcast the following graph** I've redrawn this for the sake of clarity. For the original, see http://cloudfront.mediamatters.org/static /images/item/fbn-cavuto-20120731-bushexpire.jpg.

30 **Discontinuity in vertical or horizontal axis** Spirer, Spirer, & Jaffe, op. cit., pp. 82–84.

33–34 **Choosing the proper scale and axis** Example from Spirer, Spirer, & Jaffe, op. cit., p. 78.

35 **Many things change at a constant rate** Spirer, Spirer, & Jaffe, op. cit., p. 78.

36 **life expectancy of smokers versus nonsmokers at age twenty-five** These data taken from Jha, P., et al. (2013). 21st-century hazards of smoking and benefits of cessation in the United States. *New England Journal of Medicine, 368*(4), 341–350, Figure 2A for women. Survival probabilities were scaled from the National Health Interview Survey to the U.S. rates of death from all causes at these ages for 2004 with adjustment for differences in age, educational level, alcohol consumption, and adiposity (body-mass

index). I'm grateful to Prabhat Jha for her correspondence about interpreting this.

This form of presentation is based on that of Wainer, H. (1997). *Visual Revelations: Graphical Tales of Fate and Deception from Napoleon Bonaparte to Ross Perot.* New York: Copernicus/Springer-Verlag.

38 **expenditures per public school student and those students' scores on the SAT** This example from Wainer, H. (1997). *Visual Revelations: Graphical Tales of Fate and Deception from Napoleon Bonaparte to Ross Perot.* New York: Copernicus/Springer-Verlag, p. 93. The original appeared in *Forbes* (May 14, 1990).

Of course, there are other variables. Are the spending increases reported in actual or inflation-adjusted dollars? Was the time frame 1980–88 chosen to make that point, and would a different time frame make a different point?

39 **The correlation also provides a good estimate** There is some controversy about whether to use *r* or *r-squared*. For the defense of *r*, see: D'Andrade, R., & Dart, J. (1990). The interpretation of r versus r² or why percent of variance accounted for is a poor measure of size of effect. *Journal of Quantitative Anthropology, 2,* 47–59.

Ozer, D. J. (1985). Correlation and the coefficient of determination. *Psychological Bulletin, 97*(2), 307–315.

40 **services provided by the organization Planned Parenthood** Roth, Z. (2015, Sept. 29). Congressman uses misleading graph to smear Planned Parenthood. msnbc.com.

Politifact explored this issue further, examining the data between the endpoints and furnishing additional contextual information to go along with the usual graph-centered criticism. See https://perma.cc/P8NY-YP49.

47 **presentation on iPhone sales** http://qz.com/122921/the-chart-tim-cook-doesnt-want-you-to-see/; http://www.tekrevue.com/tim-cook-trying-prove-meaningless-chart/.

49 **feature spurious co-occurrences** http://www.tylervigen.com/spurious-correlations.

51 **Randall Munroe in his Internet cartoon *xkcd*** https://xkcd.com/552/.

52 **visual system is pitted against your logical system** This example is based on one in Huff, ibid.

53 **Any model of consumer behavior on a website** This is nearly a direct quote from De Veaux, R. D., & Hand, D. J. (2005). How to lie with bad data. *Statistical Science, 20*(3), 231–238, p. 232.

54 **Colgate's biggest competitor was named nearly as often** I thank my student Vivian Gu for this example.

Derbyshire, D. (2007, Jan. 17). Colgate gets the brush off for "misleading" ads. *The Telegraph*. Retrieved from http://www.telegraph.co.uk/news/uknews/1539715/Colgate-gets-the-brush-off-for-misleading-ads.html.

54 **C-SPAN advertises that they are "available"** http://www.c-span.org/about/history/.

54 **doesn't mean that even one person is watching** Nielsen reports that Americans, on average, receive 189 channels but watch only 17 of them. http://www.nielsen.com/us/en/insights/news/2014/changing-channels-americans-view-just-17-channels-despite-record-number-to-choose-from.html.

55 **water use in the city of Rancho Santa Fe** Boxall, B. (2014, Dec. 2). Rancho Santa Fe ranked as state's largest residential water hog. *Los Angeles Times*. http://www.latimes.com/local/california/la-me-water-rancho-20141202-story.html.

Lovett, I. (2014, Nov. 29). "Where grass is greener, a push to share drought's burden." *New York Times*. http://www.nytimes.com/2014/11/30/us/where-grass-is-greener-a-push-to-share-droughts-burden.html.

56 **flying is actually safer now** http://www.flightsafety.org; Grant, K. B. (2014, Dec. 30). Deadly year for flying—but safer than ever. http://www.cnbc.com/id/102301598.

58 **Newton's law of cooling** For an initial temperature of 155 degrees Fahrenheit, the formula is

$$f(t) = 80e^{-0.08t} + 75.$$

61 **C-SPAN is available in 100 million homes** Bedard, P. (2010, June 22). "Brian Lamb: C-SPAN now reaches 100 million homes." *U.S. News & World Report*. www.usnews.com/news/blogs/washington-whispers/2010/06/22/brian-lamb-c-span-now-reaches-100-million-homes. Retrieved November 22, 2010.

61 **90 percent of the population is within twenty-five miles** Based on Huff, op. cit., p. 48.

62 **3,482 active-duty U.S. military personnel who died in 2010** https://www.cbo.gov/sites/default/Files/113th-congress-2013-2014/workingpaper/49837-Casualties_WorkingPaper-2014-08.pdf.

62 **total of 1,431,000 people in the military** http://www.census.gov/compendia/statab/2012/tables/12s0511.pdf.

62 **death rate in 2010** http://www.cdc.gov/nchs/fastats/deaths.htm.

62 **general population of the United States includes** Based on an example from Huff, op. cit., p. 83.

63 **increase in the number of doctors** I thank my student Alexandra Ghelerter for this example. Barnett, A. (1994). How numbers are tricking you. Retrieved from http://www.sandiego.edu/statpage/barnett.htm.

65   **nuances often tell a story** This is Best's term.

66   **there are *six* different indexes** Davidson, A. (2015, July 1). The economy's
      missing metrics. *New York Times Magazine.*

66   **July 2015 that the unemployment rate dropped** Shell, A. (2015, July 2).
      Wall Street weighs Fed's next move after jobs data. *USA Today Money.*
      http://americasmarkets.usatoday.com/2015/07/02/wall-street-gets-what-it
      -wants-in-june-jobs-count/.

66   **reported the reason for the apparent drop** Schwartz, N. D. (2015, July 3).
      Jobless rate fell in June, with wages staying flat. *New York Times*, B1.

71   **batting averages for the 2015 season** Stats from http://mlb.mlb.com/stats
      /sortable.jsp#elem=[object+Object]&tab_level=child&click_text=Sortable
      +Player+hitting&game_type=%27R%27&season=2015&season_type=ANY
      &league_code=%27MLB%27&sectionType=sp&statType=hitting&page=
      1&ts=1457286793822&playerType=QUALIFIER&timeframe=.

73   **top three causes of death in 2013** http://www.cdc.gov/nchs/fastats/leading
      -causes-of-death.htm.

79   **attitudes do not seem to fall upon racial lines** This is entirely
      hypothetical.

79   **Another hurdle: You want age variability** This is from Huff, op. cit., p. 22.

81   **71 percent of *which* British?** Ibid.

81   **answer falsely just to shock the pollster** Many years ago, Chicago colum-
      nist Mike Royko encouraged readers to lie to exit pollers on Election Day
      in the hope that inaccurate data and being made to look foolish would end
      the practice of TV commentators calling the result of an election before all
      the votes were counted. I have no data on how many people lied to the exit
      pollers because of Royko's column, but the fact that exit polls are still a
      thing suggests it wasn't enough.

82   **the price you pay for not hearing from everyone** Taken from http://www
      .aapor.org/AAPORKentico/Education-Resources/For-Researchers/Poll-Survey
      -FAQ/What-is-the-Margin-of-Sampling-Error.aspx.

82   **Note that these ranges overlap** This is a good rule of thumb, but in some
      cases this quick method will be inaccurate. See Schenker, N., & Gentleman,
      J. F. (2001). On judging the significance of differences by examining the over-
      lap between confidence intervals. *American Statistician*, 55(3), 182–186.

83   **Five times out of a hundred** I'm intentionally not making a distinction
      here between frequentist and Bayesian probability estimates, a distinction
      that comes up in Part Two.

84   **Margin of error** (image) From Wikipedia.

84   **formula for calculating the margin of error** For large populations, the 95
      percent confidence interval can be estimated as ±1.96 × sqrt [p(1-p)/n]. To

obtain a 99 percent confidence interval, multiply by 2.58 instead of 1.96. Yes, the interval is *larger* when you're more confident (which should make sense; if you want to be more sure that the range you quote includes the true value, you need a larger range). For smaller populations, the formula is to first compute the standard error:

sqrt [{(Observed proportion) × [I – (Observed proportion)}/sample size]

The width of the 95 percent confidence interval then is ±2 × standard error.

For example, if you sampled fifty overpasses in a large city, you might have found that 20 percent of them needed repair. You calculate the standard error as:

sqrt [(.2 × .8)/50] = sqrt (.I6/50) = .057.

So the width of your 95 percent confidence interval is ±2 × .057 = ±.11 or ±11%. Thus the 95 percent confidence interval is that 20 percent of the overpasses in this town need repair, plus or minus 11 percent. In a news report, the reporter might say that the survey showed 20 percent of overpasses need repair, with a margin of error of 11 percent. To increase the precision of your estimate, you need to sample more. If you go to 200 overpasses (assuming you obtain the same 20 percent figure), your margin of error reduces to about six percent.

85 **this conventional explanation is wrong** Lusinchi, D. (2012). "President" Landon and the 1936 *Literary Digest* poll: were automobile and telephone owners to blame? *Social Science History, 36*(1), 23–54.

85 **An investigation uncovered serious flaws** Clement, S. (2013, June 4). Gallup explains what went wrong in 2012. *Washington Post.* https://www .washingtonpost.com/news/the-fix/wp/2013/06/04/gallup-explains-what -went-wrong-in-2012/.

http://www.gallup.com/poll/162887/gallup-2012-presidential-election -polling-review.aspx.

88 **trying to figure out what proportion of jelly beans** Taken from http:// www.ropercenter.uconn.edu/support/polling-fundamentals-total-survey -error/.

89 **what magazines people read** Elaborated from an example in Huff, op. cit., p. 16.

90 **Gleason scoring** This definition taken verbatim from http://www.cancer .gov/publications/dictionaries/cancer-terms?cdrid=45696. Accessed March 20, 2016.

91 **they had made an error in measurement** Jordans, F. (2012, Feb. 23). CERN researchers find flaw in faster-than-light measurement. *Christian Science Monitor.* http://www.csmonitor.com/Science/2012/0223/CERN-researchers -find-flaw-in-faster-than-light-measurement.

91 **1960 U.S. Census study recorded** This is from De Veaux, R. D., & Hand, D. J. (2005). How to lie with bad data. *Statistical Science, 20*(3), 231–238, p. 232. They cite Kruskal, W. (1981). Statistics in society: problems unsolved and unformulated. *Journal of the American Statistical Association, 76*(375), 505–515, and Coale, A. J., & Stephan, F. F. (1962). The case of the Indians and the teen-age widows. *Journal of the American Statistical Association, 57,* 338–347.

92 **claimed measurement error as part of their defense** Kryk, J. Patriots strike back with compelling explanations to refute deflate-gate chargers. *Ottawa Sun,* May 15, 2015. http://www.ottawasun.com/2015/05/14/pat riots-strike-back-with-compelling-explanations-to-refute-deflate-gate -chargers.

94 **statistic you encounter may not have defined homelessness** This example from Spirer, H., Spirer, L., & Jaffe, A. J. (1998). *Misused Statistics,* 2nd ed., revised and expanded. New York: Marcel Dekker, p. 16.

95 **Imagine that you've been hired by a political candidate** This example based on one in Huff, op. cit., p. 80.

95 **A newspaper reports the proportion of suicides** From Best (2005), op. cit.

97 **I'm not going to wear my seat belt because** This example comes from Best, J. (2012), and my childhood friend Kevin.

98 **the idea of symmetry and equal likelihood** The principle of symmetry can be broadly construed to include instances where outcomes are not equally likely but still prescribed, such as a trick coin that is weighted to come up heads two-thirds of the time, or a roulette wheel in which some of the troughs are wider than others.

98 **If we run the experiment on a large number of people** We could also conduct the experiment with a small number of people many times, in which case we would expect to obtain different numbers. In this case, the true probability of the drug working is going to be somewhere close to the average (the mean) of the numbers obtained in all the experiments, but it's an axiom of statistics that larger samples lead to more accurate results.

98 **Both classic and frequentist probabilities deal with** Classic probability can be thought of in two different ways: empirical and theoretical. If you're going to toss a coin or draw cards from a shuffled deck, each time you do this is like a trial in an experiment that could go on indefinitely. In theory, you could get thousands of people to toss coins and pick cards for several years and tally up the results to obtain the proportion of time that different

outcomes occur, such as "getting heads" or "getting heads three times in a row." This is an *empirically derived* probability. If you believe the coin is fair (that is, there's no manufacturing defect that causes it to come up on one side more than the other), you don't need to do the experiment, because it should come up heads half of the time (probability = .5) in the long run, and we arrive at this *theoretically*, based on the understanding that there are two equally likely outcomes. We could run a similar experiment with cards and determine empirically and theoretically that the chances of drawing a heart are one in four (probability = .25) and that the chances of drawing the four of clubs is one in fifty-two (probability ≅ .02).

99   **When a court witness testifies about the probability** Aitken, C. G. G., & Taroni, F. (2004). *Statistics and the Evaluation of Evidence for Forensic Scientists,* 2nd ed. Chicester, UK: John Wiley & Sons.

100 **In Tversky and Kahneman's experiments** Tversky, A., & Kahneman, D. (1974). Judgment under uncertainty: heuristics and biases. *Science, 185*(4157), 1124–1131.

101 **A telltale piece of evidence that this is subjective** For further discussion, and more formal treatment, see Iversen, G. R. (1984). *Bayesian Statistical Inference.* Thousand Oaks, CA: Sage, and references cited therein.

106 **the case of Sally Clark** I thank my student Alexandra Ghelerter for this example. See also Nobles, R., & Schiff, D. (2007). Misleading statistics within criminal trials. *Medicine, Science and the Law, 47*(1), 7–10.

108 **relative incidence of pneumonia** http://www.nytimes.com/health/guides /disease/pneumonia/prognosis.html.

108 **Bayes's rule to calculate a conditional probability** Bayes's rule is:

$$P(A \mid B) = \frac{P(B \mid A) \times P(A)}{P(B)}$$

111 **The probability that a woman has breast cancer** This paragraph, and this discussion, quotes nearly verbatim from Krämer, W., & Gigerenzer, G. (2005). How to confuse with statistics or: the use and misuse of conditional probabilities. *Statistical Science, 20*(3), 223–230.

112 **To make the numbers work out easily** How do you know what number to choose? Sometimes it takes trial and error. But it's also possible to figure it out. Because the probability is .8 percent, or eight people per thousand, if you chose to build a table for 1,000 women you'd end up with eight in one of the squares, and that's okay, but later on we're going to be multiplying that by 90 percent, which will give us a decimal. There's nothing wrong with that, it's just less convenient for most people to work with decimals. Increasing our population by an order of magnitude to 100 gives us all

whole numbers, but then we're looking at larger numbers than we need. It doesn't really matter because all we're looking for is probabilities and we'll be dividing one number by another anyway for the result.

116 **If you read that more automobile accidents occur at seven p.m.** Still confused? If there were eight times as many cars on the road at seven p.m. than at seven a.m., the *raw* number of accidents could be higher at seven p.m., but that does not necessarily mean that the *proportion* of accidents to cars is greater. And *that* is the relevant statistic to you: not how many accidents happen at seven p.m., but how many accidents occur per thousand cars on the road. This latter formulation quantifies your risk. This example is modified from one in Huff, op. cit., p. 78, and discussed by Krämer & Gigerenzer (2005).

118 **90 percent of doctors treated the two** Cited in Spirer, Spirer, & Jaffe, op. cit., p. 197: Thompson, W. C., & Schumann, E. L. (1987). Interpretation of statistical evidence in criminal trials, *Law and Human Behavior, 11*(167).

118 **One surgeon persuaded ninety women** From Spirer, Spirer, & Jaffe, op. cit., first reported in Hastie, R., & Dawes, R. M. (1988). *Rational Choice in an Uncertain World.* New York: Harcourt Brace Jovanovich.

The original report of the surgeon's work appeared in McGee, G. (1979, Feb. 6). Breast surgery before cancer. *Ann Arbor News,* p. B1 (reprinted from the *Bay City News*).

119 **As sociologist Joel Best says** Best, op. cit., p. 184.

PART TWO: EVALUATING WORDS

125 **Steve Jobs delayed treatment for his pancreatic cancer** Swaine, J. (2011, Oct. 21). Steve Jobs "regretted trying to beat cancer with alternative medicine for so long." http://www.telegraph.co.uk/technology/apple/8841347 /Steve-Jobs-regretted-trying-to-beat-cancer-with-alternative-medicine-for -so-long.html.

125 **an article in *Forbes* that claims** Rees, N. (2009, Aug. 13). Policing word abuse. *Forbes.* http://www.forbes.com/2009/08/12/nigel-rees-misquotes -opinions-rees.html.

125 ***Respectfully Quoted,* a dictionary of quotations** Platt, S., ed. (1989). *Respectfully Quoted.* Washington, D.C.: Library of Congress. For sale by the Supt. of Docs., USGPO.

125 **That book reports various formulations** Billings, J. (1874). *Everybody's Friend, or Josh Billing's Encyclopedia and Proverbial Philosophy of Wit and Humor.* Hartford, CT: American Publishing Company.

129 **humans had twenty-four pairs of chromosomes instead of twenty-three** Gartler, S. M. (2006). The chromosome number in humans: a brief history.

Nature Reviews Genetics, 7, 655–660. http://www.nature.com/scitable/con tent/The-chromosome-number-in-humans-a-brief-15575. Glass, B. (1990). *Theophilus Shickel Painter.* Washington, D.C.: National Academy of Sciences. http://www.nasonline.org/publications/biographical-memoirs /memoir-pdfs/painter-theophilus-shickel.pdf. Retrieved November 6, 2015.

133 **If people in the arts and humanities have won a prize** Paul Simon, Stevie Wonder, and Joni Mitchell can be considered experts in songwriting. Although they do not hold university positions, university scholars have written books and articles about them, and Mr. Simon and Mr. Wonder were recognized by the president of the United States with Kennedy Center Honors, reserved for individuals who have made great contributions to performing arts. Ms. Mitchell received an honorary doctorate of music and won the Polaris Music Prize.

136 **Some people, including Noam Chomsky, have argued** Chomsky, N. (2015, May 25). The *New York Times* is pure propaganda. *Salon.* http://www.salon .com/2015/05/25/noam_chomsky_the_new_york_times_is_pure_proganda _partner/.

    Achbar, M., Symansky, A., & Wintonick, P. (Producers), and Achbar, M., & Wintonick, P. (Directors). (1992). *Manufacturing Consent: Noam Chomsky and the Media* (Motion picture). USA: BuyIndies.com Inc. and Zeitgeist Films. https://www.youtube.com/watch?v=BsiBl2CaDFg.

136 **A 2011 fake tweet** Melendez, E. D. (2013, Feb. 1). Twitter stock market hoax draws attention of regulators. http://www.huffingtonpost.com/2013/02/01 /twitter-stock-market-hoax_n_2601753.html; http://www.forbes.com /forbes/welcome/.

136 **"The use of false rumors and news reports"** Farrell, M. (2015, July 14). Twitter shares hit by takeover hoax. *Wall Street Journal.* http://www.wsj .com/articles/twitter-shares-hit-by-takeover-hoax-1436918683.

137 **Jonathan Capehart wrote a story** (2010, Sept. 7). *Washington Post* writer falls for fake congressman Twitter account. *Huffington Post*, updated Sept. 7, 2010. http://www.huffingtonpost.com/2010/09/07/washington-post -writer-fa_n_707132.html; http://voices.washingtonpost.com/postpartisan /2010/09/obama_deficits_and_the_ditch.html.

138 **Who is behind it?** This is taken verbatim from *The Organized Mind.* Levitin, D. J. (2014). *The Organized Mind.* New York: Dutton.

139 **2014 congressional race for Florida's thirteenth district** Leary, A. (2014, Feb. 4). Misleading GOP website took donation meant for Alex Sink. *Tampa Bay Times.* http://www.tampabay.com/news/politics/stateroundup /misleading-gop-website-took-donation-meant-for-alex-sink/2164138.

141 **A court case ruled that Degil** Pink, D. (2013). Deceiving domain names not allowed. Wickwire Holm. http://www.wickwireholm.com/Portals/0 /newsletter/BLU%20Newsletter%20-%20January%202013%20-%20Deceiving

%20Domain%20Names%20Not%20Allowed.pdf; Bonni, S. (2014, June 24). The tort of domain name passing off. *Charity Law Bulletin* 342, Carters Professional Corporation. http://www.carters.ca/pub/bulletin/charity /2014/chylb342.htm.

141 **vendor operated the website GetCanadaDrugs.com** https://www.canada drugs.com/; https://www.getcanadadrugs.com/ (no longer available); Naud, M. (n.d.). Registered trade-mark canadadrugs.com found deceptively misdescriptive. ROBIC. http://www.robic.ca/admin/pdf/682/293 .045E-MNA2007.pdf.

141 **MartinLutherKing.org contains is a shameful assortment** The inflammatory quote from the website comes from the book by Taylor Branch, *Pillar of Fire*, but the author notes that he did not hear the tapes himself, he took them from three FBI agents who reported them to him.

141 **Stormfront, a white-supremacy, neo-Nazi hate group.** Sources which identify Stormfront as the Internet's "first hate site" include:

Levin, B. (2003). "Cyberhate: A legal and historical analysis of extremists' use of computer networks in America," in Perry, B., ed., *Hate and Bias Crime: A Reader*. New York: Routledge, p. 363.

Ryan, N. (2004). *Into a World of Hate: A Journey Among the Extreme Right*. New York: Routledge, p. 80.

Samuels, S. (1997). "Is the Holocaust unique?," in Rosenbaum, Alan S., ed., *Is the Holocaust Unique?: Perspectives on Comparative Genocide*. Boulder, CO: Westview Press, p. 218.

Bolaffi, G.; et al., eds (2002). *Dictionary of Race, Ethnicity and Culture*. Thousand Oaks, CA: Sage Publications, p. 254.

145 **Energy-drink company Red Bull paid** O'Reilly, L. (2014, Oct. 8). Red Bull will pay $10 to customers disappointed the drink didn't actually give them "wings." http://www.businessinsider.com/red-bull-settles-false-advertising -lawsuit-for-13-million-2014-10.

145 **Target agreed to pay $3.9 million** Associated Press. (2015, Feb. 11). Target agrees to pay $3.9 million in false-advertising lawsuit. http://journal record.com/2015/02/11/target-agrees-to-pay-3-9-million-in-false-advertising -lawsuit-law/.

146 **Kellogg's paid $4 million to settle** Federal Trade Commission. (2009, April 20). Kellogg settles FTC charges that ads for Frosted Mini-Wheats were false [Press release]. https://www.ftc.gov/news-events/press-releases /2009/04/kellogg-settles-ftc-charges-ads-frosted-mini-wheats-were-false.

147 **The *Washington Post* also runs a fact-checking site** https://www.washing tonpost.com/news/fact-checker/.

148 **Politifact summarized its findings** Carroll, L. (2015, Nov. 22). Fact-checking Trump's claim that thousands in New Jersey cheered when

World Trade Center tumbled. http://www.politifact.com/truth-o-meter/
statements/2015/nov/22/donald-trump/fact-checking-trumps-claim
-thousands-new-jersey-ch/.

148 **only one grandparent was born abroad** Sanders, K. (2015, April 16). In
Iowa, Hillary Clinton claims "all my grandparents" came to the U.S. from
foreign countries. http://www.politifact.com/truth-o-meter/statements
/2015/apr/16/hillary-clinton/hillary-clinton-flubs-familys-immigration
-history-/.

150 **322,000,000** The population of the United States, as of this writing. http://
www.census.gov/popclock/.

150 **For coronary heart disease** American Heart Association (2015). AHA Statistical Update. *Circulation*, 131, p. 434–441. I thank McGill University
Librarians Robin Canuel and Genevieve Gore for help in finding these
statistics.

154 **In 1968, Will and Ariel Durant wrote** Durant, W., & Durant, A. (1968).
The Lessons of History. New York: Simon & Schuster.

155 **The FBI announced in 2015** Federal Bureau of Investigation (2015, April
20). FBI testimony on microscopic hair analysis contained errors in at least
90 percent of cases in ongoing review [Press release]. https://www.fbi.gov
/news/pressrel/press-releases/fbi-testimony-on-microscopic-hair-analysis
-contained-errors-in-at-least-90-percent-of-cases-in-ongoing-review.

155 **Without these pieces of information** Aitken, C. G. G., & Taroni, F. (2004).
Statistics and the Evaluation of Evidence for Forensic Scientists, 2nd ed.
Chicester, UK: John Wiley & Sons, p. 95, citing Friedman, R. D. (1996).
Assessing Evidence. *Michigan Law Review, 94*(6), 1810–1838.

156 **In one case in the U.K.** R v. Dennis John Adams, (1996) 2 Cr App R, 467;
And Aitken, C. (2003). Statistical techniques and their role in evidence
interpretation. In Payne-James, J., Busuttil, A., & Smock, W., eds., *Forensic
Medicine: Clinical and Pathological Aspects*. Cambridge, UK: Cambridge
University Press.

156 **the *New York Times* described a mysterious formation** Blumenthal, R.
(2015, Nov. 3). Built by the ancients, seen from space. *New York Times*,
p. D2.

159 **How much *more* productive and creative might she have been** I thank
Stephen Kosslyn for sharing a version of this example with me.

159 **Two twins were separated at birth** Grimes, W. (2015, Nov. 13). Jack Yufe,
a Jew whose twin was a nazi, dies at 82. *New York Times*, p. B8.

159 **They were reunited twenty-one years later** Much of this is taken verbatim
from Grimes (2015), op. cit.

160 **A statistician or behavioral geneticist would say** Dr. Jeffrey Mogil, personal communication.

163   **if you ask a hundred people in a room** The formula is $1 - (1 - 1/2^5)^{100}$.

164   **Paul McCartney and Dick Clark** I thank Ron Mann for this observation.

165   *Larger samples more accurately reflect* Note that in a large sample, you are more likely to find an anomalous (outlier) observation than in a small sample, but when looking at the *mean*, the mean of the large sample is far more likely to reflect the true state of the world (because there are so many more observations that can swamp the anomalous one).

166   **if the study was on the incidence of preterm births** Krämer, W., & Gigerenzer, G. (2005). How to confuse with statistics or: the use and misuse of conditional probabilities. *Statistical Science, 20*(3), 223–230. See also Centers for Disease Control and Prevention. Preterm birth. http://www.cdc .gov/reproductivehealth/maternalinfanthealth/pretermbirth.htm.

166   **Consider a street game in which a hat** Krämer, W., & Gigerenzer, G. (2005). Technically, they note, this is an incorrect enumeration of simple events in a Laplacian experiment in the subpopulation composed of the remaining possibilities.

167   **similar mistakes were made by mathematical philosopher Gottfried Wilhelm Leibniz** Ibid.

168   **Counterknowledge, a term coined by** Thompson, D. (2008). *Counterknowledge: How We Surrendered to Conspiracy Theories, Quack Medicine, Bogus Science, and Fake History.* New York: W. W. Norton, p. 1.

168   **Damian Thompson tells the story** Thompson, D. (2008), op. cit.

170   **Shot on a consumer-grade camera** Trask, R. B. (1996). *Photographic Memory: The Kennedy Assassination, November 22, 1963.* Dallas: Sixth Floor Museum, p. 5.

170   **A *handful* of unexplained anomalies** Thanks to Michael Shermer for this.

170   **The difference between a false theory** This is a direct quote from Thompson, D. (2008), op. cit.

173   **As Damian Thompson notes** Thompson, D. (2008), op. cit., p. 17. The previous two sentences are from pp. 16–17.

173   **die each year of stomach cancer** National Cancer Institute. SEER stat fact sheets: stomach cancer. http://seer.cancer.gov/statfacts/html/stomach .html.

173   **than of unintentional drowning** Centers for Disease Control and Prevention. Unintentional drowning: get the facts. http://www.cdc.gov/Homeand RecreationalSafety/Water-Safety/waterinjuries-factsheet.html.

174   **A front-page headline in the *Times* (U.K.)** Smyth, C. (2015, Feb. 4). "Half of all Britons will get cancer during their lifetime." *Times.* www.thetimes .co.uk/tto/health/news/article4343681.ece.

175   **Cancer Research UK (CRUK) reports that** Boseley, S. (2015, Feb. 3). Half of people in Britain born after 1960 will get cancer, study shows. *Guardian.*

175 **Heart disease is better controlled** Griffiths, C., & Brock, A. (2003). Twentieth century mortality trends in England and Wales. *Health Statistics Quarterly, 18*(2), 5–17.

177 **This is based on reports by a variety** http://www.nrdc.org/water/drinking/qbw.asp; http://www.mayoclinic.org/healthy-lifestyle/nutrition-and-healthy-eating/expert-answers/tap-vs-bottled-water/faq-20058017; http://www.consumerreports.org/cro/news/2009/07/is-tap-water-safer-than-bottled/index.htm; http://news.nationalgeographic.com/news/2010/03/100310/why-tap-water-is-better/; http://abcnews.go.com/Business/study-bottled-water-safer-tap-water/story?id=87558; http://www.telegraph.co.uk/news/health/news/9775158/Bottled-water-not-as-safe-as-tap-variety.html.

177 **In New York City; Montreal; Flint, Michigan; and many other older cities** Stockton, N. (2016, Jan. 29). Here's how hard it will be to unpoison Flint's water. *Wired.* http://www.wired.com/2016/01/heres-how-hard-it-will-be-to-unpoison-flints-water/.

PART THREE: EVALUATING THE WORLD

179 **Nature permits us to calculate only probabilities** Feynman, R. P. (1985). *QED: The Strange Theory of Light and Matter.* Princeton, NJ: Princeton University Press.

182 **A case of fraud occurred in 2015** Reardon, S. (2015, July 1). US vaccine researcher sentenced to prison for fraud. *Nature, 523*, p. 138.

182 **controversy about whether the measles, mumps, and rubella MMR vaccine causes autism** Wakefield, A. J., et al. (1998, Feb. 28). RETRACTED: Ileal-lymphoid-nodular hyperplasia, non-specific colitis, and pervasive developmental disorder in children. *Lancet, 351*(9103), 637–641. http://www.thelancet.com/journals/lancet/article/PIIS0140-6736(97)11096-0/abstract.

Burns, J. F. (2010, May 25). British medical council bars doctor who linked vaccine with autism. *New York Times*, p. A4. http://www.nytimes.com/2010/05/25/health/policy/25autism.html.

Associated Press (2011, Jan. 6). Study linking vaccine to autism is called fraud. *New York Times.* http://query.nytimes.com/gst/fullpage.html?res=9C02E7DC1E3BF935A35752C0A9679D8B63.

Rao, T. S., & Andrade, C. (2011). The MMR vaccine and autism: sensation, refutation, retraction, and fraud. *Indian Journal of Psychiatry, 53*(2), 95–96.

192 **For example, Holmes concludes that** From Thompson, S. (2010). The blind banker. *Sherlock* (TV series, first aired October 31, 2010).

194 **The germ theory of disease** I first learned about this story from Hempel, C. (1966). *Philosophy of Natural Science.* Englewood Cliffs, NJ: Prentice-Hall.

199 **To fill out the rest of the table** I'm using the 168-hour week (7 days × 24 hours a day) to account for thoughts you might have while dreaming, and people who might call and wake you from a sound sleep. Of course, one could subtract out eight hours of sleep per night (or whatever) and then use only the 112 hours of wakefulness to come up with a different probability, but it doesn't change the conclusion.

202 **less safe mode of travel** In retrospect this switch was foolish, at the time, but it may have been the rational thing to do. Four hijacked, suicide planes at once was unprecedented in aviation history. When confronted with a big change in the world, often the best thing to do is to think Bayesian: update your understanding, stop relying on the old statistics, and seek alternatives.

202 **conclude that air travel** Based on Huff, op. cit., p. 79.

202 **There were not nearly as many flights in 1960** See, for example, Iolan, C., Patterson, T., & Johnson, A. (2014, July 28). Is 2014 the deadliest year for flights? Not even close. CNN. http://www.cnn.com/interactive/2014/07 /travel/aviation-data/; and Evershed, N. (2015, March 24). Aircraft accident rates at historic low despite high-profile plane crashes. *Guardian*. http:// www.theguardian.com/world/datablog/2014/dec/29/aircraft-accident-rates -at-historic-low-despite-high-profile-plane-crashes.

203 **An FBI page reports that** http://www.fbi.gov/about-us/cjis/ucr/crime-in -the.u.s/2011/crime-in-the-u.s.-2011/clearances.

204 **All home robberies in a neighborhood** Image from http://contactglenda .com/wp-content/uploads/2011/08/robbers-decamp.png.

206 **In a famous psychology experiment** Nisbett, R. E., & Valins, S. (1972). Perceiving the causes of one's own behavior. In Kanouse, D. E., et al., eds. *Attribution: perceiving the causes of behavior.* Morristown, NJ: General Learning Press, pp. 63–78.

And, Valins, S. (2007). Persistent effects of information about internal reactions: ineffectiveness of debriefing. In London, H., & Nisbett, R. E., eds. *Thought and Feeling: the cognitive alteration of feeling states.* Chicago, IL: Aldine Transaction.

207 **Between 1990 and 2010, six times as many** What is causing the increase in autism prevalence. *Autism Speaks Official Blog,* Oct. 22, 2010. http:// blog.autismspeaks.org/2010/10/22/got-questions-answers-to-your-questions -from-the-autism-speaks%E2%80%99-science-staff-2/.

208 **The majority of the rise** Ibid.

208 **the Internet to guide your thinking on why autism** Suresh, A. (2015, Oct. 13). Autism increase mystery solved: no, it's not vaccines, GMOs, glyphosate— or organic foods. Genetic Literacy Project. http://www.geneticliteracyproject

.org/2015/10/13/autism-increase-mystery-solved-no-its-not-vaccines-gmos
-glyphosate-or-organic-foods/.

208 **She also couches her argument** Kase, A. (2015, May 11). MIT scientist
uncovers link between glyphosate, GMOs and the autism epidemic. *Reset
.me*. http://reset.me/story/mit-scientist-uncovers-link-between-glyphosate
-gmos-and-the-autism-epidemic/.

209 **no evidence that thimerosal was linked to autism** Honda, H., Shimizu,
Y., & Rutter, M. (2005). No effect of MMR withdrawal on the incidence of
autism: a total population study. *Journal of Child Psychology and Psychiatry, 46*(6), 572–579. http://1796kotok.com/pdfs/MMR_withdrawal.pdf, and
many other sources.

Reardon, S. (2015). US vaccine researcher sentenced to prison for fraud.
Nature, 523(7559), p. 138.

211 **as we know, there are known knowns** Defense.gov News Transcript: DoD
News Briefing—Secretary Rumsfeld and Gen. Myers, United States
Department of Defense (defense.gov).

214 **We can clarify Secretary Rumsfeld's four possibilities with a fourfold
table** I thank Morris Olitsky for this.

217 **One of the cornerstone principles of forensic science** Inman, K., & Rudin,
N. (2002). The origin of evidence. *Forensic Science International, 126*(1),
11–16.

Inman, K., & Rudin, N. (2000). *Principles and Practice of Criminalistics:
the profession of forensic science*. Boca Raton, FL: CRC Press.

218 **Suppose a criminal breaks into the stables** I'm basing this section on the
discussion found in Aitken, C. G. G., & Taroni, F. (2004). *Statistics and the
Evaluation of Evidence for Forensic Scientists*, 2nd ed. Chicester, UK: John
Wiley & Sons, pp. 1–2, and using their setup and terminology.

218 **take literally the assumption in the American legal system** Aitken,
C. G. G., & Taroni, F. (2004), op. cit.

221 *the prosecutor's fallacy* Thompson, W. C.; Shumann, E. L. (1987). Interpretation of statistical evidence in criminal trials: the prosecutor's fallacy and
the defense attorney's fallacy. *Law and Human Behavior 2*(3), 167–187.

230 **The quality of the photographs is high** Hasselblad.com. https://www
.hq.nasa.gov/alsj/a11/a11-hass.html; http://www.wired.com/2013/07
/apollo-hasselblad/.

231 **It's been claimed that the chances of life forming on Earth** Estimates
include 1 x 10^{390}. http://evolutionfaq.com/articles/probability-life. See also
Dreamer, D. (2009, April 30). Calculating the odds that life could begin by
chance. *Science 2.0*. http://www.science20.com/stars_planets_life
/calculating_odds_life_could_begin_chance.

231 **In a TED talk with more than 10 million views** https://www.ted.com
/talks/david_blaine_how_i_held_my_breath_for_17_min?language=en.

232 **there are more than 5,000 TED-branded events** Bruno Guissani, Curator
of TEDGlobal Conference, personal communication, September 28, 2015.

232 **Fox television reported his ice-block demonstration** https://www.you
tube.com/watch?v=U6Em2OhvEJY.

233 **Out of a sense of ethics** Glenn Falkenstein, personal communication,
October 25, 2007.

234 **There was even a peer-reviewed paper** Korbonits, M., Blaine, D., Elia, M.,
& Powell-Tuck, J. (2005). Refeeding David Blaine—studies after a 44-day
fast. *New England Journal of Medicine, 353*(21), 2306–2307.

234 **The current editor of the journal searched** J. Drazen, MD, email communication, December 20, 2015.

234 **The lead author on the article told me in an email** M. Korbonits, MD,
email communication, December 25, 2015.

234 **a physician** *did* **monitor Blaine throughout the fast** Jackson, J. M., et al.
(2006). Macro- and micronutrient losses and nutritional status resulting from
44 days of total fasting in a non-obese man. *Nutrition, 22*(9), 889–897.

237 **Blaine's record was broken in 2012** http://www.guinnessworldrecords.com
/world-records/24135-longest-time-breath-held-voluntarily-male; Greno-
ble, R. (2012, Nov. 16). Breath-holding world record: Stig Severinsen stays
under water for 22 minutes (Video), *Huffington Post.* http://www.huffing
tonpost.com/2012/11/16/breath-world-record-stig-severinsen_n_2144734
.html.

240 **An article in the** *Dallas Observer* Liner, E. (2012, Jan. 13). Want to know
how David Blaine does that stuff? (Don't hold your breath). http://www
.dallasobserver.com/arts/want-to-know-how-david-blaine-does-that-stuff
-dont-hold-your-breath-7083351.

240 **preparation for the breath holding for an article in the** *New York Times*
Tierney, J. (2008, April 22). This time, he'll be left breathless. *New York
Times*, p. F1.

240 **the** *Oprah* **appearance in his blog** Tierney, J. (2008). David Blaine sets
breath-holding record. http://tierneylab.blogs.nytimes.com/2008/04/30
/david-blaine-sets-breath-holding-record.

241 **Tierney writes, "I was there"** John Tierney, email correspondence, Janu-
ary 13 and 18, 2016.

245 **Eventually all of the elements between 1 and 118** Netburn, D. (2016, Jan.
4). It's official: four super-heavy elements to be added to the periodic table.
http://www.latimes.com/science/sciencenow/la-sci-sn-new-elements-20160104
-story.html.

247   **Here's Harrison Prosper, describing this plot** Prosper, H. B. (2012, July 10). International Society for Bayesian Analysis. http://bayesian.org/forums /news/3648.

248   **Louis Lyons explains "The Higgs"** Lyons, L. (2012, July 11). http://bayesian .org/forums/news/3648.

248   **Although CERN officials announced in 2012** In articles on the Higgs boson, you may encounter reference to the 5-sigma standard of proof. Five-sigma refers to the level of probability that the scientists agreed upon before conducting the experiments—the chance of their misinterpreting the experiments had to have a confidence interval within five standard deviations (5-sigma), or 0.0000005 (recall earlier we talked about 95 and 99 percent confidence intervals—this is a confidence interval of 99.99995 percent). See http://blogs.scientificamerican.com/observations/five -sigmawhats-that/.

248   **Prosper says, "Given that the search"** Prosper, H. B. (2012, July 10). http:// bayesian.org/forums/news/3648.

248   **Physicist Mads Toudal Frandsen adds** (2014, Nov. 7). Maybe it wasn't the Higgs particle after all. Phys.org. http://phys.org/news/2014-11-wasnt-higgs -particle.html.

249   **Joseph Lykken, a physicist and director of the Fermi National Accelerator Laboratory** http://phys.org/news/2014-11-wasnt-higgs-particle.html.

249   **as *Wired* writer Signe Brewster says** http://www.wired.com/2015/11 /physicists-are-desperate-to-be-wrong-about-the-higgs-boson/.

250   **physicist Nima Arkani-Hamed told the *New York Times*** Overbye, D. (2015, Dec. 16). Physicists in Europe find tantalizing hints of a mysterious new particle. *New York Times,* p. A16.

CONCLUSION: DISCOVERING YOUR OWN

252   **A lot of things that should be scientific** These ideas and their phrasing come from: Frum, D. (2015). Talk delivered at the Colleges Ontario Higher Education Summit, November 16, 2015, Toronto, ON.

APPENDIX: APPLICATION OF BAYES'S RULE

255   **To compute Bayes's rule** Iversen, G. R. (1984). *Bayesian Statistical Inference.* Quantitative Applications in the Social Sciences, vol. 43. Thousand Oaks, CA: Sage.

ACKNOWLEDGMENTS

The inspiration for this book came from Darrell Huff's *How to Lie with Statistics*, a book that I've read several times and appreciate more with each reading. I was also a huge fan of Joel Best's *Damned Lies and Statistics*, and Charles Wheelan's *Naked Statistics*. I owe all three authors a great debt for their humor, wisdom, and insight, and I hope this book will take its place alongside theirs for anyone who wants to improve their understanding of critical thinking.

My agent, Sarah Chalfant at the Wylie Agency, is a dream: warm, attentive, supportive, and indefatigable. I feel privileged to work with her and her colleagues at TWA, Rebecca Nagel, Stephanie Derbyshire, Alba Ziegler-Bailey, and Celia Kokoris.

I am thankful to everyone at Dutton/Penguin Random House. Stephen Morrow has been my editor through four books and has made each of them incalculably better ($P < .01$). His guidance and support have been valuable. Thanks to Adam O'Brien, LeeAnn Pemberton, and Susan Schwartz. Hats off to Ben Sevier, Amanda Walker, and Christine Ball for the numerous things they do to help books reach readers who want to read them. Becky Maines was a wonderful more-than-a-copy-editor, whose breadth and depth of knowledge and clarifying additions I very much enjoyed.

ACKNOWLEDGMENTS

I'm grateful to the following for helpful discussions and comments on drafts of this manuscript: Joe Austerweil, Heather Bortfeld, Lew Goldberg, and Jeffrey Mogil. For help with specific passages, I'm indebted to David Eidelman, Charles Fuller, Charles Gale, Scott Grafton, Prabhat Jha, Jeffrey Kimball, Howie Klein, Joseph Lawrence, Gretchen Lieb, Mike McGuire, Regina Nuzzo, Jim O'Donnell, James Randi, Jasper Rine, John Tierney, and the many colleagues of mine of the American Statistical Association who helped proof the book and review the examples, especially Timothy Armistead, Edward K. Cheng, Gregg Gascon, Edward Gracely, Crystal S. Langlais, Stan Lazic, Dominic Lusinchi, Wendy Miervaldis, David P. Nichols, Morris Olitsky, and Carla Zijlemaker. My students in McGill's honors and independent research seminar helped provide some of the examples and clarified my thinking. Karle-Philip Zamor, as he has through four books, helped me enormously to prepare figures and solve all manner of technical problems, always with good cheer and great skill. Lindsay Fleming, my office assistant, has helped me to structure my time and keep my focus and assisted with the end notes, index, proofreading, fact-checking, and many other details of the book (and thanks to Eliot, Grace, Lua, and Kennis Fleming for sharing her time with me).

INDEX

Note: Page numbers in *italics* refer to illustrations.

"access" (term usage), 61
accuracy of numbers, 60–61
actors and product promotion, 135
Adams, Dennis, 156
advertising, 54, 145–46
air travel, 56, 201–2
albinism, 166
Aldrin, Buzz, 229–231
Alexa.com, 143
alternative explanations, 152–167
  for ancient earthworks of
    Kazakhstan, 156–58
  and cherry-picking, 161
  and control groups, 158–160
  in court cases, 155–56
  scientists' consideration of, 249
  and selective small samples, 164–66
  and selective windowing, 161–64
amalgamating, 64–72, 74
anti-evolutionism, 170
anti-refugee sentiments, 203
anti-science bias, 252
anti-skepticism bias, 252–53
Apple, 47, *48*
arguments in science, 193–94
Arkani-Hamed, Nima, 250
Armstrong, Neil, 229–231
Associated Press (AP), 66
association, persuasion by, 176–77, 210
atoms, 245

authority, 129, 135. *See also* experts
  and expertise
autism and vaccines, 182, *207*, 207–10
automobile accidents, 116–17, 202,
  273n116
averages, 11–25
  and bimodal distribution, 17–18, *18*
  combining samples from disparate
    populations, 17
  and common fallacies, 18–20
  mean, 11–13, 17, 18–19, 21, 265n18
  median, 11, 12–13, 17, 18, 19, 265n18
  mode, 11, 13, 17, 19
  and range, 17
  and shifting baselines, 20–24, *22*, *23*
  and skewed distributions, 20–21, 24, *25*
axes, 26–42
  choosing the scale and, 33–35, *34*, *35*
  discontinuity in, 30–33, *31*, *32*,
    72–73, *73*
  double Y-axis, 36–42, *37*, *38*, *44*,
    44–45
  truncated vertical, 28–30, *29*
  unlabeled, 26–28, *27*, *28*, *40*, 41

baselines for comparison, 7, 21–24, *22*,
  23, 56
batting averages, 71–72, *72*, *73*
Bayesian reasoning, 99–102, 216–221,
  222, 223–29

Bayes's rule, 108–9, 217–221, 255, 272n108
Beall, Jeffrey, 144
belief perseverance, 205–7, 210
Best, Joel, 119
Billings, Josh, 125–28
bimodal distribution, 17–18, *18*
bin widths, unequal, 71–72
birth rates, *67*, 67–71, *68*, *69*, *70*, *71*, 151, 154
Blaine, David, 231–245
    breath-holding demonstration, 231, 236–245
    fasting demonstration, 231, 234–36
    ice-block demonstration, 231, 232–33
    needle-in-hand demonstration, 234
    TED talk of, 231, 232, 236, 244
bosons, 246. *See also* Higgs boson
bottled water, 176–77
breast cancer, 111–14, *115*, *118*, 118
breath-holding demonstration of Blaine, 236–245
Brewster, Signe, 249–250
British Medical Journal, 210
Bush, George W., 19, *29*, 29–30, *30*

Cage, Nicolas, *49*, 49–50
California, water consumption in, 55
cancer
    breast cancer, 111–14, *115*, 118, *118*
    and perception of risks, 174–75
Capehart, Jonathan, 137
car accidents, 116–17, 202, 273n116
Catholic Church, 154
causation, 48–51
celebrities, reported deaths of, 136
cell phones, 5, 61
Census of 1960 (U.S.), 91–92
CERN (European Organization for Nuclear Research), 90–91, 152, 248, 250
Chaffetz, Jason, 40–41
cherry-picking bias, 161, 165

Chomsky, Noam, 136
citations, 149
Clark, Dick, 164, 251
Clark, Sally, 106–7, 134
classic probabilities, 98–99, 271n98
Clinton, Hillary, 148
CNN, 202–3
coincidence, 49–50
collecting numbers. *See* data collection
college size, 24
combining probabilities, 102–3, 218
comparisons, 61–73
conditional probabilities, 104–8
    asymmetry in, 115–19
    and Bayes's rule, 108–9, 272n108
    in court cases, 155
    and deductive arguments, 186
    visualizing, with fourfold tables, 108–11, 118–19
confidence intervals, 82–84, *84*, 269n84, 282n248
confounding factors, 212
conspiracies and conspiratorial thinking, 168–170
Consumer Price Index (CPI), 94
Consumer Reports, 252–53
contrapositive statements, 187, 189
control groups, 158–160
converse error (fallacy of affirming the consequent), 186–87, 189
Cook, Tim, 47, *48*
correlation
    and causation, 48–51
    definition of, 39
    illusory, 198–200, *199*, 209–10
    in school funding and SAT scores, *38*, *39*, 39–40
counterknowledge, 168–177
    and conspiratorial thinking, 168–170
    defined, 168
    on Internet, 251, 254
    and journalism, 171–73
    and moon-landing denial, 230, 231

in Orwell's *1984*, 251
and perception of risks, 173–75
and persuasion by association,
176–77, 210
court cases, 16, 117, 124, 155–56
coverage errors in sampling, 87–88
crime rates, *31*, 31–33, *32*, *33*, 173
critical thinking
cultivation of, 222
and emotions, 224–29
enterprises emphasizing, 252–53
and the Internet, 252–53
and known unknowns, 250
ongoing process of, 253
and the scientific method, 252
and the scientific revolution, 181
C-SPAN watchers, 54
cum hoc, ergo propter hoc, 49
cumulative sales graphs, 46–47, *47*
currency of information, 146

Dallas Observer, 240, 242
data collection, 75–96
and definitions, 93–95
and fraud, 181–82
and measurement errors, 90–93
and participation biases, 86–88
and reporting biases, 88–89
and sampling, 76–85
and sampling biases, 85–86
and selective windowing, 162
standardization issues in, 90
for unknowable/unverifiable results,
95–96
David and Goliath (Gladwell), 159
death. *See* mortality and death
deductive reasoning, 183–190, 192,
193, 194–97
definitions in data collection, 93–95
Democrats, 62–63, 139–140
Dentec Safety Specialists and Degil
Safety Products, 140–41
Dey, Dimitriy, 157–58
discredited information, 146–48

distribution
bimodal distribution, 17–18, *18*
skewed, 20–21, 24
doctors and employment trends, 63
domains of websites, 137–38, 141
drowning deaths, *49*, 49–50, 173
Durant, Will and Ariel, 154

ecological fallacy, 18–19
education, purpose of, 215
Eidelman, David, 237
Einstein, Albert, 250
electrons, 246
elementary particles, 246
emotions
and critical thinking skills, 224–29
and justification of decisions, 124
and misframing risks, 202, 203, 205
employment/unemployment rates, 66
evolution, 170
exception fallacy, 18, 19–20
experimental design, 213
experts and expertise, 129–151
and citations, 149
and discredited information, 146–48
evidence-based arguments of, 129, 130
and fraud, 134
and hierarchy of information
sources, 135–37, 232
incorrect assessments of, 129, 130–31
and institutional bias, 142
narrow expression of, 134–35,
251–52
opinions shared by, 129–130
and peer-reviewed journals, 132,
135, 143–44
recognizing, 131–34, 274n133
and regulating authorities, 145–46
and reposted information from
others, 148
standards for, 130, 274n133
and terminology, 149–151
and websites, 137–141, 143
extrapolation, 58–60, *59*, *60*

fact-checking sites and organizations, 147–48
Falkenstein, Glenn, 233–34, 243–44, 245
fallacies, 198–210
    affirming the consequent (converse error), 186–87, 189
    belief perseverance, 205–7, 210
    cum hoc, ergo propter hoc, 49
    ecological fallacy, 18–19
    exception fallacy, 18, 19–20
    and framing of probabilities, 200–201
    gambler's fallacy, 166–67
    illusory correlations, 198–200, *199*, 209–10
    post hoc, ergo propter hoc, 48–49, 207–8, 210
    prosecutor's fallacy, 221
    when framing risk, 201–5
false positives/negatives, 111–14, *114*
family size, average, 21–22, *23*
far transfer, 222
fasting demonstration of Blaine, 234–36
Federal Bureau of Investigation (FBI), 155
fermions, 246
food labeling, 74
Forbes magazine, 38, 39, 66, 125
forensic science, 217–221
fourfold tables, 108–11, 118–19, 198–200, 217–221
Fox News, *10*, 10, *29*, 29–30
Fox television, 232
framing, 53–56, 155, 200–205
Frandsen, Mads Toudal, 248
fraudulent advertising claims, 145–46
fraudulent research data, 181–82
frequentist probabilities, 98–99, 271n98
Fuller, Charles, 237, 240

Galileo Galilei, 77–78
Gallup, George, 85

gambler's fallacy, 166–67
genetics, 159–160
Get What You Pay For (Keppel), 7
Gladwell, Malcolm, 159
Gleason tumor grading, 90
Grafton, Scott, 238–240
graphs
    choosing the scale and axes for, 33–35, *34*, *35*
    cumulative sales graphs, 46–47, *47*
    with discontinuity in axes, 30–33, *31*, *32*, 72–73, *73*
    with double Y-axis, 36–42, *37*, *38*, *44*, 44–45
    misinterpretation of, 119–120
    smooth lines in, 42
    with truncated vertical axes, 28–29, *29*
    with unlabeled axes, 26–28, *27*, *28*
gravitons, 250
Guinness World Records, 238

Han, Dong-Pyou, 182
Harte, Bret, 128
Hawking, Stephen, 249
heart disease, 150–51, 175
heterogeneity, 18
hierarchy of source quality, 135–37, 232
Higgs boson, 217, 246–250, *247*, 282n248
Holmes, Sherlock, 191–92
Holocaust revisionism, 170
homelessness, measuring, 94
home prices, 33–34, *34*
home robberies, 203–5
hypotheses, 212

illegal immigrants, 89
illusory correlations, 198–200, *199*, 209–10
illustrations, deceptive, 51–52, *52*
immigrants, 203
incidence (term), 149–151
inductive reasoning, 183–84, 190–91, 192

inference, 191
inflation measures, 94
infoliteracy, x
institutional bias, 142
intelligence, 93, 134
interest rates, 7
Internet and websites, 137–141
   and anti-skepticism bias, 252–53
   assessing authority of, 143
   citing research, 149
   counterknowledge on, 251, 254
   domains of, 137–38, 141
   misinformation on, 253
   and reposted information, 148
interpolation, 58–59
investments, 24, *25*

Jenkins, James, 21
Jobs, Steve, 125
joint probabilities, 102–3
Jolly, David, 139–140
journalism
   breaking news mode in, 171–72
   and counterknowledge, 171–73
   crime reporting in, 173
   and hierarchy of source quality,
      135–36
   reliability of, 145
   scientific investigation mode in,
      171, 172
   and selective windowing, 162–63
Journal of Management Development, 6
journals, reputable, 143–44

Kahneman, Daniel, 100
Kaysing, Bill, 229, 231
Kazakhstan, ancient earthworks of,
   156–58
Kennedy, John F., 169–170
Keppel, Dan, 7
Kerry, John, 19
King, Martin Luther, 141, 275n141
known knowns, 211–15, *214*
known unknowns, 211–15, *214*, 250

The Lancet, 210
Landon, Alf, 85
law of large numbers, 165
learning, role of mistakes in, 249
Leibniz, Gottfried Wilhelm, 167
Lieb, Gretchen, 126–28
life expectancy, 20–21, 228, 266
Literary Digest, 85
Locard, Edmond, 217–18
logarithmic scales, 35
Los Angeles Times, 55
lying to pollsters, 88–89, 269n81
Lykken, Joseph, 249, 250
Lyons, Louis, 248

magicians, 153, 232, 233–34, 242–45.
   See also Blaine, David
Malthus, Thomas, 151
margin of error, 82, 83–84, *84*, 269n84
MartinLutherKing.org, 141, 275n141
McCartney, Paul, 164, 251
mean, 11–13, 17, 18–19, 21, 265
measurement errors in data collection,
   90–93
median, 11, 12–13, 17, 18, 19, 265
medicine
   and decision making, 111–15, 117–18
   and effect of treatments on life
      expectancy, 228
   employment trends in, 63
   and incidence vs. prevalence,
      149–151
   and mortality, 150–51
   and pathology, 224
Mendeleev, Dmitri, 245
meta-analyses, 142, 183, 217
military service, 62
misinformation, xi, 168, 254, 263. *See
   also* counterknowledge
mistakes and learning
   opportunities, 249
mode, 11, 13, 17, 19
modus ponens statements, 186, 188
moon-landing denial, 229–231

mortality and death
causation vs. correlation in,
49–50, *49*
and cause-of-death statistics, 73–74
and conditional probabilities, 106–7
crude death rate, 151
and medical advances of
Semmelweis, 194–97
mortality term, 150–51
and perception of risk, 173–75
and plausibility assessments, 6
reported by media, 136
and risk-framing fallacies, 202–3
and sampling errors, 62, 63
and shifting baselines, 56
and unknowable/unverifiable
claims, 95–96
motorcycle accidents, 63
Mozart effect, 158
multiple sclerosis (MS), 150
multiraciality, 89
Munroe, Randall, 51

neutrons, 246
New England Journal of Medicine, 234
New England Patriots, 92
New Jersey, welfare mothers in, 8–9
New York Times
on ancient earthworks of
Kazakhstan, 156–57
on Blaine's breath-holding
demonstration, 240–41, 242
on Higgs boson research, 250
statistics misreported by, 8–9
on unemployment rate, 66
and verification of information,
135–36
on water consumption in
California, 55
1984 (Orwell), 251
Nixon, Tom, 243–44
non-response errors in sampling, 87
notation of logical statements, 187–89
Nutrition, 234–35

Orwell, George, 251
outliers, 12–13
The Oxford Dictionary of Quotations
(Knowles, ed.), 126–27

Paris attacks of November 13, 2015,
202–3
participation biases, 86–88
pathology, 224
pattern detection, 198
peer review, 132, 135, 143–44, 145
percentages and percentage points, 7
periodic table, 245–46
Personal Consumption Expenditures
(PCE), 94
persuasion by association, 176–77, 210
pie charts, 9–10, *10*
placebos, 158–59
plane crashes, 56, 201–2
Planned Parenthood, *40*, 40–42, *41*,
44–45
plausibility, assessing, 3–10
pneumonia, 108
political parties, 62–63
Politifact.com, 147–48
Pollack, Louis, 6
Popoff, Peter, 241
population
population size, 83
rate of natural increase of (RNI), 151
sample sizes for surveys of, 165
post hoc, ergo propter hoc, 48–49,
207–8, 210
Potkin, Ralph, 241
precision vs. accuracy of numbers,
60–61
predictions, 154–55
presidential elections, 19, 85, 91,
147–48
prevalence (term), 149–151
probabilities, 97–120
Bayesian, 99–102, 216–221
classic, 98–99, 271n98
combining, 102–3, 218

and counterknowledge, 170
deductive logic of, 183
of events informed by other events,
    103–4
framing, 200–201
frequentist, 98–99, 271n98
in medicine, 224
for recurring, replicable events,
    98–99
and statistical literacy, 166–67
subjective, 99–102
visualizing, with fourfold tables,
    108–11, 118–19, 198–200
See also conditional probabilities
prosecutor's fallacy, 221
Prosper, Harrison, 247, 248
protons, 246, 247
psychics, 153
public-opinion polls, 76, 99
public policy, 54, 64–65
publishing, 135, 145

quarks, 246
quota sampling, 81
quotations, 125–28

racial identities, 89
Randi, James, 153, 241, 244
random sampling, 79–81, 85, 163
range and averages, 17
recycling, 54–55
Rees, Nigel, 125–26
refugees, 203
regret, 229
regulating authorities, 145–46
reporting bias in data collection, 88–89
representative sampling, 76–80,
    161–64
Republicans, 62–63, 139–140
Respectfully Quoted (Platt), 125, 126–27
Richards, Cecile, 42
risk
    framing of, 201–5
    perception of, 173–75

Rivera, Geraldo, 173
Rocketdyne, 230–31
Rogers, Will, 126, 127, 128
Roosevelt, Franklin D., 85
Rumsfeld, Donald, 211, 213–15

Sacramento Bee, 173
salespeople, 203–4
samples from disparate populations, 17
sampling, 76–85
    biases in, 85–86
    and confidence intervals, 82–84, *84*,
        269n84
    and margin of error, 82, 83–84, *84*,
        269n84
    methods of, 62–63
    and probabilities, 104
    quota, 81
    random, 79–81, 85
    representative, 76–80
    sample size, 67, 83, *84*, 271n98
    selective small samples, 164–66
    stratified, 79–81
SAT scores and school funding, *38*,
    38–40, *39*
scenario planning, 154–55
Science, 6
science and the scientific method,
    181–197
    and abductive reasoning, 191–92
    and alternative explanations, 249
    and anti-science bias, 252
    arguments in, 193–94
    and critical thinking skills, 252
    and deductive reasoning, 183–190,
        192, 193, 194–97
    and experimental design, 213
    and hypotheses, 212
    and inductive reasoning, 183–84,
        190–91, 192
    and medical advances of
        Semmelweis, 194–97
    and meta-analyses, 142, 183, 217
    myths about, 182–83

science and the scientific method (*cont.*)
  and notation of logical statements,
    187–89
  and scientific revolution, 181
  value of basic research in, 252
selective windowing, 161–64
Semmelweis, Ignaz, 194–97
Seneff, Stephanie, 208–9
September 11, 2001, terrorist attacks,
    169, 202
Severinsen, Stig, 237
sexual behaviors of teens and preteens,
    64–65
Shaw, Gordon, 134
shifting baselines, 20–24, *22, 23*
Shockley, William, 134
SIDS (sudden infant death syndrome),
    106–7
significance, statistical, 57–58
Sink, Alex, *139,* 139–140, *140*
Slovic, Paul, 173
smoking deaths, *36,* 36–37
Snopes.com, 252
sources, hierarchy of, 135–37, 232
sporting events and probabilities, 101
standardization issues in data
    collection, 90
Standard Model of Particle Physics,
    246–250
standard of proof, 248, 282n248
statistical literacy, 166–67
Stephanopoulos, George, 147–48
stratified random sampling, 79–81
subdividing, 73–74
subjective probabilities, 99–102
syllogisms, 185–86

Tampa Bay Times, 147
TED talks, 231, 232, 236, 244
telephone sampling, 62–63, 85–86
terrorism, 169, 202–3
third factor x explanation of
    correlations, 50–51

Tierney, John, 240–42, 244
Time magazine, 5, 61
Times (U.K.), 174–75
TMZ.com, 136, 137, 232
toilets, access to, 5, 61
transportation and safety
    considerations, 201–2
trials, legal, 16, 117, 124, 155–56
Trump, Donald, 147–48
Tuten, Terrell, 21
Tversky, Amos, 100
Twain, Mark, 125, 127, 128

United Nations, 61
unknown unknowns, 211–15, *214*
USA Today, 66
U.S. Bureau of Labor Statistics, 66, 94
U.S. Centers for Disease Control,
    73–74
U.S. News & World Report, 62, 63
U.S. Steel, 16

vaccines and autism, 182, 207–10
verification of information,
    147–48
veterinary science, 223–29
Vigen, Tyler, 49–50

Wakefield, Andrew, 182, 210
Wall Street Journal, 136–37, 171
Washington Post, 147–48, 171
water quality, 176–77
water usage in California, 55
weather patterns, 103–4
websites. *See* Internet and websites
Weiss, Hans, 6
World Health Organization
    (WHO), 209
Wright brothers, 131

xkcd (Munroe), 51

Zander, Benjamin, 249